Bayesian Cost–Effectiveness Analysis of Medical Treatments

Chapman & Hall/CRC Biostatistics Series

Shein-Chung Chow, Duke University School of Medicine
Byron Jones, Novartis Pharma AG
Jen-pei Liu, National Taiwan University
Karl E. Peace, Georgia Southern University
Bruce W. Turnbull, Cornell University

Recently Published Titles

Self-Controlled Case Series Studies: A Modelling Guide with R
Paddy Farrington, Heather Whitaker, Yonas Ghebremichael Weldeselassie

Bayesian Methods for Repeated Measures
Lyle D. Broemeling

Modern Adaptive Randomized Clinical Trials: Statistical and Practical Aspects
Oleksandr Sverdlov

Medical Product Safety Evaluation: Biological Models and Statistical Methods
Jie Chen, Joseph Heyse, Tze Leung Lai

Statistical Methods for Survival Trial Design
With Applications to Cancer Clinical Trials Using R
Jianrong Wu

Bayesian Applications in Pharmaceutical Development
Satrajit Roychoudhury, Soumi Lahiri

Platform Trials in Drug Development: Umbrella Trials and Basket Trials
Zoran Antonjevic and Robert Beckman

Innovative Strategies, Statistical Solutions and Simulations for Modern Clinical Trials
Mark Chang, John Balser, Robin Bliss and Jim Roach

Bayesian Cost-effectiveness Analysis of Medical Treatments
Elías Moreno, Francisco José Vázquez-Polo and Miguel Ángel Negrín-Hernández

For more information about this series, please visit: https://www.crcpress.com/go/biostats

Bayesian Cost–Effectiveness Analysis of Medical Treatments

Elías Moreno

Department of Statistics, University of Granada, Granada, Spain

Francisco José Vázquez-Polo
Miguel Ángel Negrín-Hernández

Department of Quantitative Methods, University of Las Palmas
de Gran Canaria, Las Palmas de Gran Canaria, Spain

CRC Press
Taylor & Francis Group
Boca Raton London New York

CRC Press is an imprint of the
Taylor & Francis Group, an **informa** business

A CHAPMAN & HALL BOOK

CRC Press
Taylor & Francis Group
6000 Broken Sound Parkway NW, Suite 300
Boca Raton, FL 33487-2742

First issued in paperback 2020

© 2019 by Taylor & Francis Group, LLC
CRC Press is an imprint of Taylor & Francis Group, an Informa business

No claim to original U.S. Government works

ISBN 13: 978-1-138-73173-8 (hbk)
ISBN 13: 978-0-367-73187-8 (hbk)

**Visit the Taylor & Francis Web site at
http://www.taylorandfrancis.com**

**and the CRC Press Web site at
http://www.crcpress.com**

To Nines for her supports along the time (Elías Moreno)

Contents

Preface

Cost–effectiveness analysis is the heading of a set of statistical methodologies for evaluating medical treatments. It is required that the evaluation be based on the effectiveness and cost of the treatments. As the effectiveness and cost change across the patient population to which a given treatment is applied, it is assumed that they are random variables for which a statistical model has to be proposed. In this setting, the aim of the analysis is that of choosing the "optimal" treatment in a specified set of alternative treatments for a given disease whose cost and effectiveness follow specified statistical models.

The interest in cost–effectiveness analysis has grown in the last three decades, and the literature on the topic is spread out in scientific journals that include the *European Journal of Health Economics, European Journal of Operational Research, Health Economics, Journal of Health Economics, Medical Decision Making, Pharmacoeconomics* and *Value in Health*. A few books, including Briggs et al. (2006), Willan and Briggs (2006), Baio (2012) and Drummond et al. (2015), put together the advancements in different aspects of cost–effectiveness analysis to date.

This book provides the basis of statistical decision theory and methods required for the cost–effectiveness analysis and is aimed at students of health economics with a solid background in statistics, applied statisticians, researchers in health economics, and health managers.

We realized that treatment comparison problems can be included as a particular case of the general statistical decision problem, and that statistical decision theory provides the theoretical justification to the traditional cost–effectiveness analysis methodology. Furthermore, decision theory suggests improvements of the analysis. For instance, the specification of alternative utility functions to the one typically used for the traditional incremental net benefit analysis is immediately suggested. Moreover, the identification in cost–effectiveness analysis of the elements of a statistical decision problem such as the decision space, the space of rewards, and the utility function helps in understanding the conditions we are assuming when choosing optimal treatments. As a consequence, the main motivation of this book was the formulation of cost–effectiveness analysis as a statistical decision problem and the application of the well–established statistical decision methodology to it.

The first three chapters of the book are dedicated to the exposition of some basic ideas in cost–effectiveness analysis, in Bayesian and frequentist statistical inference for parametric models, and in statistical decision theory.

Chapter 1 presents an overview of traditional methods for the economic evaluation of medical treatments which mainly apply to pairwise treatment comparisons.

Chapter 2 summarizes the basic notions of Bayesian and frequentist parametric estimation, hypothesis testing, and prediction for a few useful parametric statistical models. We remark that hypothesis testing is treated as a particular case of the Bayesian model selection problem. The reason is that the results of model selection given by the Bayesian and the frequentist approaches differ, and the Bayesian approach considerably improves the frequentist, as is well–documented in the literature of Bayesian statistics.

Chapter 3 provides a summary of the elements, principles, and results of general statistical decision theory. A central role is played by the utility function as the tool for ordering rewards of alternative decisions, and in particular, rewards of alternative medical treatments for a given disease.

Difficulties and challenges in the application of statistical decision theory to cost–effectiveness analysis are discussed in the last three chapters. The formulation of the cost–effectiveness analysis as a statistical decision problem, the specification of the reward distributions, and the utility functions implicitly or explicitly used in the existing literature are presented in Chapter 4. The parametric models frequently used for costs and effectiveness are also given. These models will be utilized for illustrating notions, methods, and case studies.

The realistic case where samples of costs and effectiveness of a given treatment are collected in different health–care centers are presented in Chapter 5. The rationale is that even when the protocol for the application of the treatment is the same in the different health–care centers, it seems reasonable to accept that a certain degree of heterogeneity of the samples across centers is always present. It is seen that ignoring the between–sample heterogeneity and pretending that they are homogeneous might give serious misleading results. To deal with heterogeneous models in cost–effectiveness analysis requires specific statistical techniques such as clustering and meta-analysis. These statistical techniques are formulated here from the Bayesian viewpoint, and presented in Chapter 5.

In the last chapter, Chapter 6, cost–effectiveness analysis for subgroups of patients is considered. This subgroup analysis appears when a potential set of covariates of patients under treatments are included in the statistical model of cost and effectiveness. Given that patient subgroups are defined by the set of covariates, a previous step in subgroup analysis is that of eliminating from the model noninfluential covariates. This is an old and basic problem in regression models known as the variable selection problem. The Bayesian variable selection approach is presented as a model selection problem, and prior distributions for models and for model parameters are discussed.

In the book we present, whenever possible, optimal treatments derived using both the frequentist and the Bayesian approach to the underlying statistical inference. However, model selection problems such as variable selection or clustering are studied using only Bayesian procedures because of their methodological simplicity and excellent sampling behavior. Moreover, although the original formulation of meta–analysis with the well–known random effect models is a mixing of frequentist and Bayesian approaches, a fully Bayesian meta–analysis is here proposed.

Computational difficulties inherent in the Bayesian model selection problem sometimes require numerical methods. When it is the case, the codes utilized for finding a numerical solution are written in `Mathematica` software, and they are available upon request to the authors.

Las Palmas de Gran Canaria, Canary Islands, Spain
September 2018

Elías Moreno
Francisco José Vázquez-Polo
Miguel Ángel Negrín-Hernández

Acknowledgments

We are grateful to Professors Guido Consonni, Carles Cuadras, Francisco J. Girón, Luis R. Perichi, Ludovico Piccinato and Walter Racugno for suggesting corrections that improved the draft of the book. Nevertheless, any mistake, imprecision, or typo is our exclusive responsibility.

The MINECO and Ministerio de Educación (Spain) provided financial support through the grants ECO2017–85577–P and MTM2014–55372–P.

Authors

Elías Moreno is Professor of Statistics and Operational Research at the University of Granada, Spain, Corresponding Member of the Royal Academy of Sciences of Spain, and elect member of ISI.

Francisco José Vázquez-Polo is Professor of Mathematics and Bayesian Methods at the University of Las Palmas de Gran Canaria, and Head of the Department of Quantitative Methods.

Miguel Ángel Negrín-Hernández is Senior Lecturer in the Department of Quantitative Methods at the ULPGC. His main research topics are Bayesian methods applied to Health Economics, economic evaluation and cost-effectiveness analysis, meta-analysis and equity in the provision of healthcare services.

1

Health economics evaluation

1.1 Introduction

In the field of economics, health economics is one of the areas in which research has recently had an intensive development. In this context, one of the major concerns of researchers is the comparison between medical treatments or technologies. Comparing the effectiveness of different treatments is not enough for medical treatment decision making. If we choose only the effectiveness as the measure of goodness of a treatment, we are accepting an unlimited capacity of resources for health, and the reality is that health resources are limited and effectiveness comes at a price. Weinstein and Stason (1977) stated that for a level of available resources, society must maximize the total aggregate of health benefits. Therefore, it is necessary to search for a methodology for adding the cost to the effectiveness, as well as a relationship between the cost and effectiveness that allows us to compare treatments.

As control over health expenditure has increased over the last thirty years, the term cost–effectiveness has gained in popularity. This increasing focus on cost–effectiveness analysis of new or existing treatments has been led by the development of health technology assessment (HTA) agencies, such as the National Institute for Health and Care Excellence (NICE) in the United Kingdom, which seeks to provide guidelines for health–care providers and decision makers about which treatments should be covered by the National Health Systems.

Most developed countries have developed HTA agencies in recent years to inform policy making. Regional and national HTA agencies offer recommendations on medicines and other health technologies that can be financed or reimbursed by the health system of a state or region. Among their functions we can highlight that of providing relevant information about the safety, efficacy, outcomes, effectiveness, cost and cost–effectiveness, as well as social, legal, ethical, and political impacts of a health–care technology. Recently, they are gaining a strong influence on patient access to new medicines, mainly due to increasing pressure on health budgets (Ciani and Jommi, 2014).

In Europe, since 2008, the European Medicines Agency has been working closely with HTA bodies in different member states, as well as with the European Network for Health Technology Assessment, with the objective of generating relevant data for regulators, HTA bodies, and other interested parties.

In the United States, the federal government has provided financial support for HTA since the early 1970s. The US Office of Technology Assessment, the Medicare Coverage Division with the Centers for Medicare and Medicaid Services, and the Agency for Healthcare Research and Quality are some of the federal institutions that undertake or fund cost or cost–effectiveness analyses of medical technologies and interventions (Luce and Brown, 1995; Sullivan et al., 2009). In other countries, such as Australia or Canada, pharmaceutical companies are required to submit their products to cost–effectiveness analysis (Henry, 1992; Lee and McCarron, 2006; Hayley, 2009). Most of the HTA agencies belong to the International Network of Medical Technology Assessment Agencies, which promotes exchange and collaboration among different evaluation agencies.

The development of this area has aroused the interest of researchers working on the statistical and methodological aspects of the decision making process in the comparison of treatments. The cost–effectiveness analysis research increases every year. To quantify this evolution, it may be helpful to show the number of articles published in MEDLINE,[1] the main database of medical literature in the world, from 1980 to 2017 (Figure 1.1). From this figure we observe an exponentially increasing number of publications between those years.

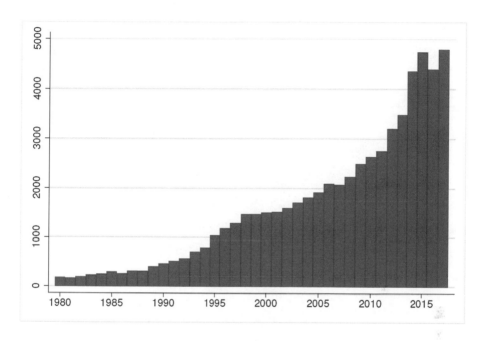

FIGURE 1.1

Number of references on economic evaluation in MEDLINE from 1980 to 2017.

In this chapter we give a historical summary of the types of economic evaluation of health technologies, data sources, tools for cost–effectiveness analysis, and a brief introduction to the Bayesian approach of cost–effectiveness analysis.

---

[1]Search strategy: (((("cost effectiveness"[Title/Abstract]) OR "cost utility"[Title/Abstract]) OR "cost benefit"[Title/Abstract]) OR "economic evaluation"[Title/Abstract]) (https://www.ncbi.nlm.nih.gov/pubmed/)

1.2   Conventional types of economic evaluation

The economic evaluation of health technologies for their comparison is based on the outcome and cost of the technologies (usually, medical treatments). Depending on how we measure the outcome of the technology, we can find three main types of methodologies for economic evaluation: cost–benefit analysis, cost–effectiveness analysis, and cost–utility analysis (Palmer et al., 1999). A succinct explanation of them is given below.

- Cost–benefit analysis.

  In this analysis, the cost and effect of the technologies are measured in commensurate terms, usually monetary (Mishan, 1988; Hutton, 1992; Robinson, 1993). The criterion for choosing the most appropriate program is simple: their benefit is greater than their cost.

  This analysis has the advantage that it allows comparison of technologies with different measures of effectiveness. A disadvantage is that it is very difficult to transform health units into monetary units. The dependence of this type of analysis on the monetary valuation of health and the methods used for its estimation means that this technique is less and less used in the evaluation of health technologies.

- Cost–effectiveness analysis.

  This analysis is an economic evaluation technique in which two or more health technologies are pairwise compared in terms of natural units of effectiveness, such as life years gained or improvements in functional status, blood pressure, cholesterol, etc. (Detsky and Naglie, 1990; Weinstein, 1990; Udvarhelyi et al., 1992; Gold et al., 1996). Costs and effectiveness are measured in non–comparable units, so one alternative will be preferred to another if it provides greater benefit at the same or lower cost, or costs less to provide the same or greater benefit. This definition does not directly address the preference for interventions which provide more effectiveness at greater cost or less effectiveness at lower cost. In that case it is necessary to set an amount that is acceptable to pay to gain a unit of effectiveness. Cost–effectiveness analysis only allows the comparison of treatments when the same outcome measure is used.

- Cost–utility analysis.

This is a label used for a special form of cost–effectiveness analysis in which effectiveness is measured in terms of Quality-Adjusted Life Years (QALYs) (Weinstein and Stason, 1977; Eisenberg, 1989). QALYs provide a composite measure of the effects of treatments that integrate the two most important dimensions of health: quantity and quality of life (Torrance, 1976; Kaplan and Bush, 1982; Torrance, 1986; Drummond et al., 1987; Torrance, 1987; Mehrez and Gafni, 1989; Torrance and Feeny, 1989; Torrance, 1995). QALYs can be interpreted as the number of healthy years equivalent to the true state of health. Cost–utility analysis offers the possibility to compare different types of interventions or health programs, considerably expanding the range of application.

As in cost–effectiveness analysis, to compare interventions which provide more effectiveness at greater cost and vice versa using cost–utility analysis, it is necessary to define a threshold that reflects the willingness to pay for a QALY. Different authors (Hirth et al., 2000; Mason et al., 2009; Pinto et al., 2009; Donalson et al., 2011; Bobinac et al., 2014) have tried to estimate the value of a QALY but there is still no consensus on this issue. As a reference, NICE has stated a threshold range of £20,000 to £30,000 per QALY gained (NICE, 2013), unless for end–of–life treatment where a threshold higher than £30,000/QALY is acceptable (Collins and Latimer, 2013).

Cost–utility analysis can be seen as a specific type of cost–effectiveness analysis in which the effectiveness is measured in terms of QALYs gained. Henceforth we will not distinguish between these methods and will use the general term cost–effectiveness analysis.

1.3   The variables of cost–effectiveness analysis

- Effectiveness

  Treatment effectiveness is one of the basic variables of a cost–effectiveness analysis. By effectiveness we understand the extent to which a treatment or health technology manages to achieve the objectives for which it was designed under ordinary circumstances (not controlled circumstances as we would see in a laboratory). Effectiveness must be distinguished from two very closely related concepts: efficacy and efficiency. Efficacy refers to the extent to which a treatment or intervention produces the desired effect under ideal conditions. Efficiency is an economic concept which relates efficacy and effectiveness to resources used. Assessment of efficiency is concerned with whether or not acceptable efficacy and effectiveness are achieved with optimal resources.

  One of the key phases of any cost–effectiveness analysis is the selection of the relevant health measure for comparing treatments. An error in choosing the right measure of effectiveness may lead to erroneous conclusions in the analysis. There is a wide variety of possible measures of effectiveness. Some are final, health–related measures of outcome, such as life–years gained or the probability of suffering a relapse. Others are expressed as intermediate outcomes, such as viral load in patients infected with Human Immunodeficiency Virus (HIV) or percentage cholesterol reduction, etc.

  There is no consensus on which measure of effectiveness should be used in each cost–effectiveness analysis. Thus, researchers should select the appropriate effectiveness measure for each analysis. In several studies (Yates, 1979; Yates et al., 1999; Negrín and Vázquez–

Polo, 2006; McCaffrey et al., 2015), the authors have considered more than one measure of effectiveness for the comparison between technologies. The most frequent procedure is to carry out an independent cost–effectiveness analysis for each measure of effectiveness. If the conclusions coincide, the decision making is easy. On the other hand, if there are contradictions on which technology is more cost–effective, the health provider should choose the measure of effectiveness that he considers most appropriate or try to find some intermediate solution. This way of solving the cost–effectiveness analysis does not take into account the correlation between the different measures of effectiveness.

Bjøner and Keiding (2004) proposed a cost–effectiveness analysis with multiple measures of effectiveness, applying Data Envelopment Analysis. Their method replaces the concept of cost–effectiveness with that of relative cost–effectiveness, where treatments are compared for different forms of aggregation of effectiveness measures. In fact, treatments that are not dominated, that is, those for which there is no other treatment that obtains more effectiveness per unit of cost for each of the measures of effectiveness considered, cannot be compared. Negrín and Vázquez–Polo (2006) proposed a Bayesian framework to carry out cost–effectiveness analysis with more than one measure of effectiveness by weighting these measures based on the willingness to pay for each effectiveness unit. More recently, McCaffrey et al. (2015) and McCaffrey and Eckermann (2016) showed how to extend cost–effectiveness analysis for multiple effects in the cost–disutility plane (Eckermann et al., 2008).

A measure of effectiveness: Quality–Adjusted Life Years.

Many studies have used the increase in life expectancy or the years of life gained as the main outcome of the effectiveness of a treatment. However, in many cases, the years of life gained is an incomplete measure of effectiveness of the treatment. In addition to the years of life gained, it is necessary to incorporate the improvement in quality of life as a relevant result in health. Therefore, it is necessary to obtain information on the health–related quality of life (HRQL) associated with each intervention. In general, the quality of life is measured continuously through a bounded index between 0 and 1, where the value 1 indicates the optimal health status. The optimum state of health is an abstract concept and has been considered by different instruments such as normal health, disease–free health, or the best state of health imaginable. The value 0 is associated with the worst state of health imaginable, which can be death or not. The comparison of treatments or interventions requires the combination of the years of life and HRQL in a single measure, giving rise to QALY (Weinstein and Stason, 1977). There are different instruments to measure the quality of life or utility associated with each state of health as time trade–off scales, standard gamble, EQ5–5D, EQ–5D–5L, SF–36, and so on (Alonso et al., 1994; Brooks, 1996; Badía et al., 1999).

Although the QALY metric represents a rigorous methodological tool for comparing treatments, there exist some limitations in its application (Dolan, 2011; Pettitt et al., 2016). Ethical and methodological issues have been widely debated in the literature (Hirskyj, 2007; Schlander and Richardson, 2009). Besides, QALYs also appear to have a limited function in some contexts, such as mental health problems (Knapp and Mangalore, 2007), elderly patients (Pinto–Prades et al., 2014) or palliative care interventions (Hughes, 2005).

The QALY framework provided a basis for the development of other health synthetic indicators, including the DALYs (Disability–Adjusted Life Years). DALYs summarize the impact on mortality and disability related to specific diseases in different communities (Sassi, 2006).

Effectiveness data sources

Randomized clinical trial is generally accepted as the most powerful design for collecting data of the effectiveness of a treatment, intervention, drug, or health technology. It is a standardized methodology that is based on well–founded scientific principles, and has been incorporated into the legal provisions on clinical evaluation of medicinal products. Thus, the Committee for Proprietary Medicinal Products, a consulting body of the European Medicines Agency, has set the standards under which clinical trials are to be conducted.

Patients are randomly assigned to the treatments, thus minimizing bias. This assignation also allows comparability of the study and control groups and provides better inferences than those obtained through an observational study. To carry out a clinical trial requires a clinical plan (a protocol), where the phases to be performed are defined. In general, two types of studies are usually distinguished, although they can be done at the same time: confirmatory studies, which try to verify the effectiveness of a technology from a very elaborate and concrete set of questions to be answered; and exploratory studies, which try to respond to a wider range of issues, and do not show the degree of accuracy that is attributable to the former.

There are some disadvantages associated with clinical trials that are summarized below:

– The delicate selection of patients.

  Patients selected for a clinical trial are, in general, highly defined. This aspect allows the conclusions to be very specific, but may make the sample not representative of the study population. For example, it may happen that those patients who agree to participate in the clinical trial are the patients who present better health outcomes. This is called the healthy volunteer effect (Hunter et al., 1987; Goodwin et al., 1988; Mandel et al., 1993).

  Another problem related to patient selection is the under–representation of minority groups. However, it is sometimes these minority groups that present the highest levels of risk.

– The difference between the results in a clinical trial and in real conditions.

The clinical trial usually reproduces the ideal conditions for the implementation of the intervention, conditions that rarely occur in ordinary practice.

– Limited time horizon.

The temporal horizon of clinical trials is usually limited so that in many cases, we obtain intermediate results. It is possible that the long–term effectiveness differs from that observed in the period of the clinical trial (Davies et al., 2013).

In addition to clinical trials, data for cost–effectiveness analysis may be provided by observational studies. Observational studies differ from clinical trials in that the investigator has no control over which patients receive treatment. There are two main types of observational studies, observational cohort studies and case–control studies. In observational cohort studies, a given sample is analyzed over time to observe the effectiveness of the treatment. They usually present a greater bias than clinical trials since each treatment can be chosen by patients with different physical or social characteristics that could have an effect on effectiveness and cost. It is therefore necessary in this type of study to control the variables that may have a relevant effect on the treatment results.

An advantage of observational studies is that the results obtained more closely resemble the true effectiveness of treatment than in clinical trials. In addition, they tend to be longer and include more patients.

A case–control study is an epidemiological, observational, analytical study, in which the patients who present with the disease, or in general a certain effect, are distinguished from those who do not (control). Once individuals are selected from each group, they are investigated as to whether they are exposed to a characteristic of interest (intervention or treatment) and the proportion of those exposed in the group of cases are compared to that of the control group. Case–control studies are useful for relatively small sample sizes (i.e., the study of rare events), require little time in execution and are relatively inexpensive compared to cohort studies.

• Cost

The analysis of the costs of an intervention includes the identification, measurement, and evaluation of all resources that are used in a given intervention. Any resource used should be considered, identified, estimated in quantitative terms, and valued monetarily. Different types of costs can be distinguished:

– Direct costs

These types of costs include the value of all goods and services that are consumed in the development of a particular intervention. Direct costs encompass all types of resources used, including time consumption by professionals, family, volunteers, and the patient. They can be directly associated with health services, such as drug costs, diagnostic tests, consultations, cost of treatment of adverse effects, hospitalization, etc., and non–health costs such as transfer to hospital, social services, or therapy.

– Indirect costs

These are also known as productivity costs. These are costs related to variations in the productive capacity of the patient, such as lost workdays. These indirect costs may be associated with loss of productivity due to illness, or loss of productivity due to death. There is a wide literature that discusses the incorporation of these costs in the economic evaluation (Ernst, 2006, and references therein).

The weight of the indirect costs in the total will depend on the technology evaluated. For example, technologies related to the treatment of influenza or certain allergies may have low direct costs, but would entail significant indirect cost savings due to the reduction of work–related casualties.

– Intangible costs

These are the costs related to the pain suffered by patients. They are not usually included in the overall costs of the technologies because of their difficult quantification.

– Transition costs

Cost–effectiveness analysis usually assumes that the treatment change is instantaneous so that the two treatments are on an equal footing. However, to implement a new treatment is likely to require some investment, for instance, some form of capital (infrastructure and equipment), personnel (training or redeployment), additional administrative complexity (data capture or new guidelines). Such investments for changing treatments are known as transition costs. As a consequence, it might be necessary to include these costs (Fenwick et al., 2008).

The valuation of costs is one of the hardest problems in the evaluation of a treatment. In relation to direct costs, for those goods and services for which there is a market, the price is accepted as an opportunity cost. However, a wide variety of health goods and services are not in a market (public services). The most common option is to measure costs in average terms, dividing the hospital budget by the number of annual stays, and eliminating the costs of resources not related to the technology to be studied. The appropriate option would be to account for all the resources used by the patient during the treatment: medicines, tests performed, days of hospitalization, staff, etc. Of course, a correct analytical accounting in public health centers would facilitate the task of calculating direct costs.

The calculation of indirect costs could be done through the mean wage values for a given cohort of patients. However, this method would present problems in evaluating the productivity of certain groups such as retirees, students, etc. (Liljas, 1998).

- Perspectives in cost–effectiveness analysis

  The perspective is the point of view from which the cost and effectiveness are assessed. It is crucial to clearly define the perspective to be used in an economic evaluation as it will have an influence on the cost and effectiveness to be evaluated.

  The societal perspective is used as the reference case, including all relevant costs and effects no matter who pays the costs or who receives them (Drummond et al., 2008). Other alternative perspectives, as that from the payer (i.e., patient) or provider (i.e., National Health System) point of view, only consider the direct costs in their analysis, neglecting to include indirect costs (Neumann, 2009).

1.4   Sources of uncertainty in cost–effectiveness analysis

The study of the uncertainty in cost–effectiveness analysis has been extensively reviewed (Manning et al., 1996; Briggs and Gray, 1999; Briggs, 2000). In this section we present different sources of uncertainty:

1. Stochastic uncertainty.

   This uncertainty refers to the natural variability that occurs among homogeneous individuals in their response to a treatment and their costs. Further evidence will not reduce this variation. It is also called *first–order uncertainty* (Briggs, 2000).

2. Heterogeneity.

   It refers to the variability of the stochastic uncertainty between subgroups of homogeneous individuals in their response to a treatment and their costs. It is due to (a) subgroups of identifiable individuals with common characteristics such as age, sex or other characteristic, or (b) unmeasured differences or latent variables. This type of uncertainty is called by Briggs

(2000) as *characteristics of the patients*. Although most of the cost–effectiveness studies compare the results of effectiveness and cost from different treatment groups by assuming that the differences between groups are not relevant, there is recent literature that incorporates covariates for defining groups showing that effectiveness and cost of a treatment may substantially differ across groups (Willan et al., 2004; Vázquez–Polo et al., 2005a; Gomes et al., 2012; Moreno et al., 2013b, 2016, among others).

Heterogeneity also arises when there are different locations in which patients receive treatment, even when the protocol employed in the different health–care centers is the same.

3. Model parameter uncertainty.

This refers to the uncertainty as to the true value of the parameters of the models utilized for the cost and the effectiveness of the treatments. The uncertainty in the estimation of the parameters of interest is also called second–order uncertainty. According to the estimation method used, the measure of the estimation uncertainty varies. The most common methods are briefly discussed below.

• *Frequentist.* In the frequentist approach to statistical inference, parameters are treated as having fixed but unknown values. From this perspective, it is not possible to associate probabilities with parameters. The uncertainty of parameter estimation is measured through *confidence intervals*. A common mistake in interpreting confidence intervals is to consider that parameters can vary, but it is the sample that can vary. A 95% confidence interval means that as the sample varies, 95% of the confidence intervals will contain the true value of the parameter. Thus, it is clear that the method for constructing the confidence interval does not depend on the observed data, and hence it cannot be interpreted as a measure of the uncertainty on the parameter estimation but a measure of the sampling uncertainty on the parameter estimator.

• *Bootstrap*. This method does not require specifying the distribution of the cost and the effectiveness. If the data include n observations, then the bootstrap analysis takes random samples from the empirical distribution for these observations. For each of the bootstrap samples, the quantity of interest is estimated. Efron and Tibshirani (1993) describe how to perform this method for ratios, such as the cost–effectiveness ratio. Some bootstrap applications in cost–effectiveness analysis are discussed in Tambour and Zethraeus (1998), Briggs et al. (1997), Gray et al. (2000), O'Hagan and Stevens (2003) and Korthals–de Bos et al. (2004).

• *Bayesian*. This methodology provides a procedure for measuring the uncertainty of parameter estimation with probabilities, since the parameters of the model are random variables with a probability distribution. This way, the Bayesian analysis offers, through the posterior distribution of the parameters, point estimates as the posterior mean, the posterior median assuming it is unique, or the global mode of the posterior distribution, and also the uncertainty of the estimation.

An idea about the uncertainty of the likelihood function for a sample is given by the likelihood sets: the smaller the likelihood sets the smaller the uncertainty of the likelihood function. An idea about the uncertainty of the parameter estimate is given by the credible sets, which also depends on the prior distribution of the parameter.

Nonparametric Bayesian analysis assumes that the distribution of the sample is random, and hence a prior distribution for the sampling distribution is necessary. This approach will not be considered in this book.

One of the first applications of the Bayesian methodology in the medical context is found in Eddy et al. (1992). Spiegelhalter et al. (1994) and Jones (1996) discuss the Bayesian approximation for statistical inference in the comparison of sanitary technologies. Parmigiani (2002) discusses Bayesian modeling in health decision making. There are many other examples in the literature of the use of the Bayesian methodology for treatment comparison (Brophy and Joseph, 1995; Heitjan, 1997; Al and Van Hout, 2000; Fryback et al., 2001; Vanness and Kim, 2002; Moreno et al., 2013b, 2014; Negrín and Vázquez–Polo, 2006; Baio, 2014; Baio et al., 2017, among others).

4. Model uncertainty.

Model uncertainty refers to the lack of knowledge regarding the appropriate model for the cost and effectiveness. It is a component of methodological uncertainty (Briggs, 2000; Negrín and Vázquez–Polo, 2008; Moreno et al., 2013b).

Typically, a set of models is proposed and the model uncertainty is measured by a probability distribution on the set.

1.5   Conventional tools for cost–effectiveness analysis

The cost–effectiveness analysis of two alternative treatments aims to combine information on both the clinical effectiveness and the costs of the treatments. A first tool used in cost–effectiveness analysis is the so-called incremental cost–effectiveness ratio. The objective of the ratio is the comparison between two alternative treatments. Generally one of the alternative treatments is a new treatment under study while the other treatment may be the one utilized until that time.

1.5.1 The incremental cost–effectiveness ratio

It is assumed that cost and effectiveness of treatment T_i, for $i = 1, 2$, follow an unknown bivariate distribution with mean (γ_i, ϵ_i). Then, the incremental cost mean is defined as the difference of the mean cost of the treatments, that is $\Delta\gamma = \gamma_1 - \gamma_2$, and the incremental effectiveness mean as $\Delta\epsilon = \epsilon_1 - \epsilon_2$. The incremental cost–effectiveness ratio (ICER) is given by

$$\text{ICER}_{12} = \frac{\Delta\gamma}{\Delta\epsilon}. \tag{1.1}$$

The subindex in the ICER denotes that the ICER is tied to treatments T_1 and T_2. The ICER_{12} can be interpreted as the increment of cost per unit of incremental effectiveness when adopting the alternative treatment T_1 instead of the control treatment T_2. Because the means are unknown, ICER_{12} should be estimated. From the frequentist approach, the ICER_{12} is estimated as

$$\widehat{\text{ICER}}_{12} = \frac{\bar{c}_1 - \bar{c}_2}{\bar{e}_1 - \bar{e}_2}, \tag{1.2}$$

where $\bar{c}_i$ and $\bar{e}_i$ represent the sample means of the cost and the effectiveness of treatment T_i, $i = 1, 2$, respectively.

The ICER_{12} has been the most used tool for decision making in cost–effectiveness analysis. However, this tool has been questioned in recent years due to difficulties in its interpretation, as well as in calculating confidence intervals for ratios. Moreover, an additional problem is that a small incremental effectiveness mean would cause the ICER_{12} to be unstable (Willan and O'Brien, 2001; O'Hagan et al., 2000).

For the representation of the ICER_{12}, the plane is divided into four quadrants where the incremental effectiveness mean ($\Delta\epsilon$) is the x–axis, and the incremental cost mean ($\Delta\gamma$) is the y–axis. The four quadrants show the different possible combinations in relation to the sign of incremental effectiveness and incremental cost mean (Figure 1.2).

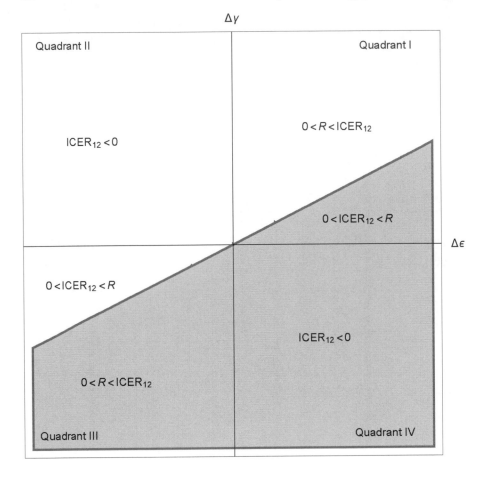

FIGURE 1.2

Cost–effectiveness plane.

Quadrants II and IV correspond to dominance of the treatments T_2 and T_1, respectively. Quadrant II indicates that treatment T_1 is more expensive and less effective than treatment T_2. In contrast, in the IV quadrant, T_1 is more effective and less expensive than T_2. In both quadrants the value of the $ICER_{12}$ ratio is negative. This is one of the limitations of $ICER_{12}$, its interpretation changes according to its sign. Thus, the same negative value may indicate dominance of the new treatment or dominance of the control treatment.

In the first quadrant, the new treatment is more effective, but also more expensive. Therefore, a subjective input is necessary to determine the preferred treatment. It is accepted that T_1 treatment is preferred when the $ICER_{12}$ is smaller than a fixed amount R. The R value represents the maximum acceptable increment of cost per unit of increment of effectiveness. Figure 1.2 shows the line $\Delta\gamma = R \cdot \Delta\epsilon$. The points below that line (in gray color) in quadrant I corresponds with $ICER_{12}$ values lower than R, which implies preference for treatment T_1. In the upper area the T_2 treatment would be preferred. In quadrant III the new treatment is less effective but less expensive than T_2. In this quadrant treatment T_1 will be preferred if the reduction increment of cost per unit of increment of effectiveness is greater than the R value. Otherwise, the preferred treatment is T_2.

Several authors have suggested the use of different threshold values depending on whether the incremental average effectiveness is positive (new treatment would improve the average effectiveness) or negative (new treatment would reduce the average effectiveness). O'Brien et al. (2002) consider the difference in willingness to accept monetary compensation for loss effectiveness with the new treatment and willingness to pay for an increase on the same average effectiveness. From that perspective the threshold represents social preferences rather than the shadow price of a fixed budget constraint.

In conclusion, for decision making, the $ICER_{12}$ estimate must be complemented with information about the quadrant in which the solution is found. For further discussion and illustrations of interpretive problems of the $ICER_{12}$ see Heitjan et al. (1999) and O'Hagan et al. (2000).

To illustrate some of the concepts introduced in this chapter we use real data from an observational study.

Example 1.1. *Data consist of a multicenter Spanish study in which various treatment regimens were compared for asymptomatic HIV patients receiving Highly Active Anti–Retroviral Therapy (Pinto et al., 2000). A cohort of several asymptomatic HIV–infected patients were* *observed under real practice and treated with two nucleoside analogues. A protease inhibitor was added and at least one nucleoside analogue was changed, following the clinical recommendations.*

The cost data correspond to direct costs (pharmaceutical, medical visit and diagnostic test costs) and the effectiveness data was expressed in QALYs. All patients used a monthly diary for six months to keep a record of resource consumption and quality of life progress. Two triple combination treatment regimens are compared. The first T_1 (d4T + ddl + IND) combines stavudine (d4T), didanosine (ddl) and indinavir (IND), and the second treatment regimen T_2 (d4T + 3TC + IND) combines stavudine (d4T), lamivudine (3TC) and indinavir (IND).

Table 1.1 provides the mean, standard deviation and sample size of the data for treatments T_1 and T_2. From Table 1.1 it follows that the sample mean of the cost and effectiveness of T_1 are greater than those of treatment T_2. The T_1 treatment is on average 160.42 euros more costly than T_2. The T_1 treatment is also more effective, 0.4024 QALYs versus 0.3958 QALYs for the T_2 treatment.

TABLE 1.1

Sample mean (standard deviation) and sample size of costs and effectiveness in Example 1.1.

	T_1	T_2
Costs (euros)	7302.70 (1702.85)	7142.28 (1568.12)
Effectiveness (QALYs)	0.4024 (0.0641)	0.3958 (0.0639)
n	95	270

For this data set, the $ICER_{12}$ is estimated from a frequentist approach by $24306.06 = 160.42/0.0066$ euros. Hence, it would be located in quadrant I. Which treatment is preferred depends on the willingness to pay for the unit of incremental effectiveness mean (R). For instance, for a standard value $R = 20000$ euros per QALY, treatment T_2 is preferred as the $ICER_{12}$ is higher than the willingness to pay R.

Confidence intervals for the ICER estimation

We note that in Example 1.1, only the means of the cost and the effectiveness have been used. Decision making based on the value of the $ICER_{12}$ should not be based solely on the sample means. Further analysis of the uncertainty associated with such an estimation is necessary. In addition to the interpretive problems presented by the $ICER_{12}$, difficulties are found in the calculation of the confidence intervals (Tambour et al., 1998). In order to determine the accuracy of the $ICER_{12}$ estimates, different techniques for computing confidence intervals have been proposed (Polsky et al., 1997), such as (i) the Box method (O'Brien et al., 1994; Wakker and Klaasen, 1995), (ii) Taylor series method (O'Brien et al., 1994; O'Hagan and Stevens, 2001, 2003), (iii) Fieller's method (Fieller, 1954; Chaudhary and Stearns, 1996; Willan and O'Brien, 1996; Laska et al., 1997; Heitjan, 2000), (iv) nonparametric bootstrapping (Chaudhary and Stearns, 1996; Briggs et al., 1997), and (v) confidence ellipses (Van Hout et al., 1994). Excluding the nonparametric bootstrap case, the remainder of the methods need to assume a probability distribution for the effectiveness and cost, usually the normal distribution.

The confidence interval for the $ICER_{12}$ estimation for any of the above methods, presents serious methodological limitations. We bring them here only for historical reasons. We use again the data from Example 1.1 in order to compute the confidence interval of the $ICER_{12}$ estimation obtained by the methods enumerated above.

Example 1.1 (continued). *In Table 1.2 we present the 95% confidence interval for $ICER_{12}$.*

We recall that the estimation of $ICER_{12}$ is 24306.06. In spite of this value, the Box method obtains negative values for both limits of the confidence interval. The negative value of the lower limit is due to a negative value of the lower limit for the incremental cost mean, and the negative value of the upper limit is due to a negative value of the lower limit for the incremental mean effectiveness. These numbers show that

TABLE 1.2

Confidence Interval for the $ICER_{12}$ in Example 1.1.

	95% Confidence Interval
Box method	$(-2221.20, -340329.28)$
Taylor series	$(-56587.19, 105025.74)$
Bootstrapping	$(-216398.08, 374254.10)$
Fieller's method	$(19581.64, 59366.00)$

the Box method does not make sense in this case. Wakker and Klaasen (1995) argue that "a reliable decision is not available and the decision should not be based on the ICER." The Taylor series and the boot-strapping (with 2000 replicates) methods obtain very wide confidence intervals. Fieller's method obtains a narrower confidence interval. Figure 1.3 shows the cost–effectiveness plane from bootstrap sampling.

1.5.2   The incremental net benefit

The incremental net benefit (INB) has been proposed in cost–effectiveness analysis as a tool for decision making that improves the ICER (Stinnett and Mullahy, 1998). The INB is defined as

$$INB_{12} = R \cdot (\epsilon_1 - \epsilon_2) - (\gamma_1 - \gamma_2) = R \cdot \Delta\epsilon - \Delta\gamma, \qquad (1.3)$$

where R is interpreted as the cost that the decision maker is willing to pay to increase the incremental effectiveness mean in one unit when using treatment T_1 instead of treatment T_2. The incremental net benefit thus defined is expressed in monetary units. In Chapter 4 we extend this notion, which is tight to the means of the cost and the effectiveness of two treatments. We will formalize the net benefit of a medical treatment, and we will reinterpret R as the utility of a unit of effectiveness. In this section we just give a brief historical development of the INB notion.

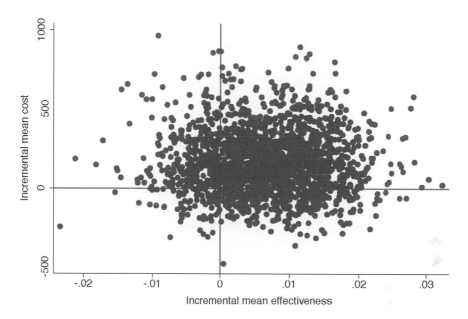

FIGURE 1.3

Points of incremental cost mean and effectiveness for 2000 bootstrap replications in Example 1.1.

The relationship between the INB and the ICER for each of the four quadrants in Figure 1.2 is as follows.

- Quadrant I: In this area, both the incremental mean cost and the incremental mean effectiveness take positive values. In this case the INB_{12} will be positive if

$$INB_{12} > 0 \iff R \cdot \Delta\epsilon - \Delta\gamma > 0 \iff R > \frac{\Delta\gamma}{\Delta\epsilon} = ICER_{12}.$$

Therefore, positive values of the INB_{12} in Quadrant I correspond to $ICER_{12}$ values below R. This implies that the increment of cost per unit of increment of effectiveness is smaller than R.

- Quadrant II: In this case, the new treatment reduces the effectiveness and increases the costs of the control treatment, therefore it is a region in which the control dominates. In this case the INB_{12} will always take negative values since R is positive.

- Quadrant III: In this quadrant, both the cost and the effectiveness increment are negative. The INB_{12} will take a positive value if

$$INB_{12} > 0 \iff R \cdot \Delta\epsilon - \Delta\gamma > 0 \iff R < \frac{\Delta\gamma}{\Delta\epsilon} = ICER_{12}.$$

Therefore, a positive INB_{12} in Quadrant III implies that the increment of cost for unit of increment is greater than R.

- Quadrant IV: In this quadrant the INB_{12} is positive. Values of the $ICER_{12}$ in this quadrant indicate that the new treatment dominates the control since it increases the effectiveness, saving costs.

It is clear that positive values of the incremental net benefit indicate a preference for the new treatment T_1 versus the control treatment T_2.

It is also obvious that if the value the decision maker is willing to pay for increasing the incremental effectiveness mean in one unit (R) coincides with the $ICER_{12}$, the incremental net benefit will be zero.

Following Stinnett and Mullahy (1998), the straightforward estimator of INB is

$$\widehat{INB}_{12} = R \cdot (\bar{e}_1 - \bar{e}_2) - (\bar{c}_1 - \bar{c}_2).$$

In contrast to the ICER, the expected value and variance of the estimator $\widehat{INB}_{12}$ are easily obtained by

$$\mathbb{E}[\widehat{INB}_{12}] = R \cdot (\epsilon_1 - \epsilon_2) - (\gamma_1 - \gamma_2),$$

$$\mathbb{V}[\widehat{INB}_{12}] = \sum_{i=1}^{2} (R^2 \sigma_{e_i}^2 + \sigma_{c_i}^2 - 2R\rho_i \sigma_{c_i} \sigma_{e_i})/n_i,$$

where $\sigma_{e_i}^2$ and $\sigma_{c_i}^2$ represent the variance of the random variables e_i and c_i, respectively, ρ_i is the correlation for e_i and c_i, and n_i is the sample size of treatment i. It is assumed that there is no correlation between treatments.

The variance of $\widehat{\text{INB}}_{12}$ is estimated as

$$\widehat{\sigma}^2_{\widehat{\text{INB}}_{12}} = \sum_{i=1}^{2} (R^2 s^2_{e_i} + s^2_{c_i} - 2Rr_i s_{c_i} s_{e_i})/n_i,$$

where $s^2_{e_i}$ and $s^2_{c_i}$ represent the sample variances; and r_i is the sample correlation for e_i and c_i.

We remark that the INB_{12} is estimated for a given value of R.

1.5.3 Cost–effectiveness acceptability curve

The cost–effectiveness acceptability curve (CEAC) was introduced by Van Hout et al. (1994) as a way to assess the uncertainty surrounding the cost–effectiveness ratio, and as an alternative to confidence intervals for the ICER estimation. The CEAC (Figure 1.4) is a function of R defined as the probability that the estimator of INB_{12} for a given R is greater than zero, that is,

$$\Pr(\widehat{\text{INB}}_{12} \geq 0) = \Pr(R \cdot (\bar{e}_1 - \bar{e}_2) - (\bar{c}_1 - \bar{c}_2) \geq 0). \qquad (1.4)$$

We note that the sample means in (1.4) are now random variables, and the probability is computed with respect to its sampling distribution. The CEAC was described by Van Hout et al. (1994) as "the probability that the ICER found in the study is acceptable" for a given R. This interpretation has been followed by several authors who have considered the CEAC as a useful tool for choosing between two alternative treatments (Van Hout et al., 1994; Fenwick et al., 2001; Fenwick and Byford, 2005).

However, the CEAC does not determine the optimal decision. The CEAC is not constructed from the mean of the distribution of the cost and the effectiveness, as is the ICER or the INB. It is easy to realize that if the distribution of the effectiveness and cost were completely known, the CEAC would not exist.

The CEAC only evaluates the uncertainty around the estimation of the INB (or ICER) and it may be useful to complement the conclusions

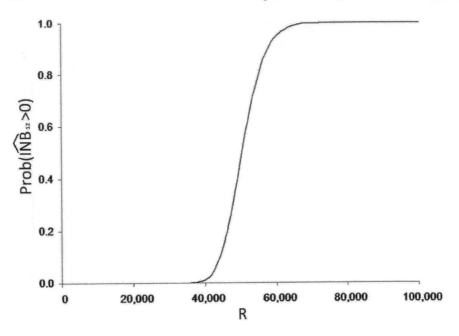

FIGURE 1.4

A typical profile of the cost–effectiveness acceptability curve.

reached by the INB, but it has no value as a tool for decision making by itself. A high probability (more than 50%) of having a positive $\widehat{INB}_{12}$ does not necessarily imply a positive estimation of the INB. This result is due to the possible asymmetry in the distribution of the INB. CEACs have been usually based on symmetric distributions and thus, a decision based on a positive mean coincides with a decision based on a probability of being positive of at least 50%. However, the high asymmetry frequently observed in the cost data does not recommend this kind of simplistic analysis.

Fenwick et al. (2004) analyzed the characteristics of the cost–effectiveness acceptability curve. A value of $R = 0$ would indicate that the decision maker is not willing to pay anything to increase effectiveness. Therefore, the CEAC will only be based on costs. The CEAC for $R = 0$ coincides with the probability of estimating a negative $\Delta\gamma$. This value would coincide with the ordinate at the origin of the cost–effectiveness acceptability curve. On the other hand, high R values indicate that the decision maker is willing to pay a lot to increase effectiveness. For R values that tend to infinity, the CEAC is based solely on which treatment is most effective. Therefore, the CEAC has a horizontal asymptote that coincides with the probability of estimating a positive $\Delta\epsilon$.

Several studies including Fenwick et al. (2004), Koerkamp et al. (2007) and Jakubczyk and Kaminski (2010) have shown the inconsistencies that can be reached when the CEAC is used as a tool for decision making. In Chapter 4 we will come back to the CEAC and will give a detailed discussion of its meaning.

1.5.4   Conventional subgroup analysis

The bulk of the literature on cost–effectiveness analysis proposes optimal treatments for the whole population without subgroup considerations. A small part of the literature assumes normal regression models for both the cost and effectiveness of the treatments, where the treatments are incorporated into the models as a binary variable and the covariates are considered for the estimation of their regression coefficients (Willke et al., 1998; Willan et al., 2004; Vázquez–Polo et al., 2005a,b).

If we denote the sample from treatment T_i as data$_i$ = $\{(c_{ij}, e_{ij}, \mathbf{x}_j), j = 1, \ldots, n_i\}$ for $i = 1, 2$ where c_{ij}, e_{ij} are the observed cost and effectiveness of patient j receiving treatment T_i, and $\mathbf{x}_j = (x_{1j}, \ldots, x_{kj})$ the vector of k covariates of patient j, a linear model for the above data that typically appears in the literature (Willan et al., 2004, and references therein) is

$$c_{ij} = \sum_{p=1}^{k} \alpha_{ip} x_{pj} + \gamma T_i + \varepsilon_i, \quad \varepsilon_i \sim \mathcal{N}(\varepsilon|0, \sigma_i^2), \qquad (1.5)$$

$$e_{ij} = \sum_{p=1}^{k} \beta_{ip} x_{pj} + \gamma' T_i + \varepsilon_i', \quad \varepsilon_i' \sim \mathcal{N}(\varepsilon'|0, \tau_i^2), \qquad (1.6)$$

where $\alpha_{ip}, \gamma, \beta_{ip}, \gamma'$ are the regression coefficients, σ_i^2, τ_i^2 the variance errors, and T_i is a $0-1$ deterministic covariate indicating the treatment. Furthermore, model (1.5–1.6) is usually restricted by the assumptions

A.1 $\alpha_{1p} = \alpha_{2p}$, and $\beta_{1p} = \beta_{2p}$ for $p = 1, \ldots, k$,

A.2 $\sigma_1 = \sigma_2$ and $\tau_1 = \tau_2$.

Under these conditions, it follows that the INB_{12} is given by

$$\text{INB}_{12} = R\gamma' - \gamma. \tag{1.7}$$

Thus, once the parameters γ and γ' are estimated, the INB provides a decision on the treatment to be chosen for each value of R.

Certainly, this *ad hoc* formulation makes the computation of the INB extremely simple but, unfortunately, equation (1.7) is true only under very stringent constraints that limit the applicability of the resulting models. Indeed, the more important restrictions are as follows:

(a)   Statistical assumptions A.1 and A.2 are unrealistic. It is not clear at all that the covariates should have exactly the same influence on the effectiveness of different treatments, and the same can be argued for the costs. In fact, examples with real data where the influential covariates for the effectiveness of a treatment change when the treatment changes are common, and the same can be observed for the cost. Further, these assumptions imply that the decision making affects the whole population as it does not depend on a specific value of the regressors, and hence it renders useless the idea of the subgroup.

(b)   The statistical model above cannot model a possible dependency between the cost c and the effectiveness e (Grieve et al., 2005). However, this dependency seems to be *a priori* a reasonable assumption. For instance, Willan et al. (2004) and Vázquez–Polo et al. (2005b) assume a multivariate normal distribution for the error terms ε_i and ε_i' according to a SURE (seemingly unrelated regression equations) model. Willke et al. (1998) and Vázquez–Polo et al. (2005b) propose an asymmetric model where the effectiveness is a covariate of the regression model for the costs.

(c) It does not cover the case where the distribution of the net benefit is not normal. However, in many applications the cost is usually assigned an asymmetric distribution, for instance a lognormal or a gamma distribution (Al and Van Hout, 2000; O'Hagan and Stevens, 2002; Fryback et al., 2001), and for these cases equation (1.7) is no longer useful.

(d) It is not clear how the above regression model (1.5–1.6) generalizes to multiple treatment comparisons, even when other treatments are included in the analysis through indicator variables. It seems that only pairwise optimizations with the reference treatment can be handled.

On the other hand, some other papers directly model the net benefit of the treatments as a linear model, where the net benefit of patient i receiving treatment j is given by $NB_{ij} = Re_{ij} - c_{ij}$ (Hoch et al., 2006; Manca et al., 2005).

Again, to keep computations simple it is assumed that for each value of R the observed net benefit of patient j, $j = 1, \ldots, n_i$, receiving treatment $i, i = 1, 2$, can be written as

$$NB_{ij} = \sum_{p=1}^{k} \eta_p x_{pj} + \delta T_i + \varepsilon, \quad \varepsilon \sim \mathcal{N}(\varepsilon | 0, \sigma^2), \qquad (1.8)$$

where the regression parameters $\eta = (\eta_1, \ldots, \eta_k)^\top$ and the variance error σ^2 are independent of the treatment. Then, it follows that the INB_{12} has the simple expression

$$\text{INB}_{12} = \delta.$$

Model (1.8) is simple but very restrictive. Restriction (a) stated above is also found here as it is assumed that the covariates have the same influence on the patients receiving different treatments. Furthermore, formulation (1.8) assumes the same unsupported constraints (c) and (d) we mentioned above.

Several authors have proposed alternative models trying to overcome the above restrictions. However, they only provide partial solutions. Manca et al. (2007) propose a bivariate hierarchical modeling for cost–effectiveness analysis using data from multinational trials. They model dependency between cost and effectiveness and carry out multiple treatment comparisons but restrictions (a) and (c) remain. Willan and Kowgier (2008) proposed a cost–effectiveness analysis with a binary measure of effectiveness and an interacting covariate. Although restriction (c) is partially avoided, restrictions (a) and (b) remain. Nixon and Thompson (2005) considered the gamma distribution for cost and effectiveness to suppress restriction (c). Further, none of these studies consider the uncertainty associated with variable selection (Negrín and Vázquez–Polo, 2008). That is, in the literature of cost–effectiveness analysis before year 2012, no attempt has been made to select a subset of influential covariates in the regression models for e_i and c_i, $i = 1, 2$, even when the elimination of non–influential covariates reduces the dimension of the model, a crucial point in regression analysis. However, a completely new formulation of the linear models in this context was given in Moreno et al. (2012).

Subgroup analysis was proposed in Willan et al. (2004), Nixon and Thompson (2005), Vázquez–Polo et al. (2005a), Vázquez–Polo et al. (2005b) and Manca et al. (2007). In all of them, subgroup analysis was carried out by the inclusion of interactions between the treatment and the subgroup in equations (1.5–1.6) or (1.8). In this case the evaluation of subgroup analyses is reduced to a statistical test for interaction. However, it is recognized that interaction terms could achieve significance by chance and it is important to acknowledge this problem to avoid spurious subgroup analyses (Pocock et al., 2002). Further, this way of analyzing subgroups is limited to discrete covariates. If we are interested in the analysis of subgroups for a continuous variable (age, weight, initial health status, etc.), discretization of the variable is then needed.

In equations (1.5), (1.6) and (1.8), treatment comparison has been converted to an estimation problem where the patient covariates and the treatments are put on an equal footing. We do not share these ideas and believe that treatment comparison is not an estimation problem but a testing problem. In Chapter 6 we follow the ideas in Moreno et al. (2012, 2013b), and as a previous step of the cost–effectiveness analysis in the presence of covariates, we propose the use of an objective Bayesian method for variable selection. We also reformulate the regression models to overcome restrictions A.1 and A.2 mentioned above and consider the treatment selection as a model selection problem.

1.6 An outline of Bayesian cost–effectiveness analysis

The frequentist approach has been the commonly used approach for estimating parameters and comparing pairs of medical treatments (Van Hout et al., 1994; Wakker and Klaasen, 1995; Willan and O'Brien, 1996; Laska et al., 1997; Stinnett and Mullahy, 1998; Tambour et al., 1998). However, the frequentist analysis has certain limitations, especially about the measurement of the uncertainty of the parameter estimation and model comparisons. The Bayesian approach gives an adequate response on how to measure these uncertainties.

Clinical research is essentially a sequential process in which every study is framed in a context of updating knowledge. The Bayesian method adjusts to this situation in a natural way since initial beliefs are sequentially modified by the new data. The possibility of incorporating prior information on the parameters of interest is an interesting feature of the Bayesian approach. The key point is that the prior knowledge about the model parameter or model is described by probability distributions.

Bayesian methodology differs from the frequentist one in the way unobservable uncertain quantities, such as model parameters or models, are considered. From a Bayesian approach, the uncertain quantities are assumed to be random variables, and thus they are described by probability distributions. Bayesian statistics generates results from the posterior distribution of model parameters or models. It is possible to produce Bayesian point estimates of model parameters or models and to measure the uncertainty of the estimation in terms of probabilities. In Chapter 2 we present a formal introduction to the Bayesian framework.

Due to the advantages of the Bayesian approach, several authors have advocated its use in cost–effectiveness analysis. Spiegelhalter et al. (2000) and O'Hagan and Luce (2003) are contributions that show how the Bayesian method applies to the evaluation of health technologies. The former article presents an analysis making use of prior information and the latter demonstrates the value of such information. The former assumes normality of the cost and the effectiveness. In a more recent work, the latter authors drop this assumption and propose a general framework (O'Hagan and Stevens, 2002). Stevens and O'Hagan (2001) made an interesting comparison between the classical and Bayesian methodologies in cost–effectiveness analysis.

In the Bayesian approach computation used to be hard, although computational methods have been developed to remedy this difficulty. We illustrate the Bayesian ICER computation using the data in Example 1.1. The data for this illustration has been taken from Pinto et al. (2000), and the priors for the model parameters have been chosen essentially for their easy implementation in the Markov Chain Monte Carlo (MCMC) machinery via `OpenBUGS` (Spiegelhalter et al., 2014); also, these priors are not the ones we would recommend.

Example 1.1 (continued). *We reanalyze the data in Example 1.1 from the Bayesian viewpoint. A bivariate normal distribution is assumed for the log of the cost and the effectiveness of treatment T_i, that is,* $\mathcal{N}((\log(c), e)^\top | (\gamma_i, \epsilon_i)^\top, \Sigma)$.

To complete the Bayesian model, a normal distribution with zero mean and variance 10000 is assumed for γ_i, and also for ϵ_i. For the covariance matrix Σ an inverse Wishart distribution is assumed with parameters I_2 and 2, where I_2 is the 2×2 identity matrix. Posterior distributions of the parameters are obtained by simulation using MCMC methods implemented in OpenBUGS software (Spiegelhalter et al., 2014). A summary of the results is shown in Table 1.3.

TABLE 1.3

Bayesian estimates in Example 1.1.

	Mean	Standard Deviation	95% Credible Interval
γ_1	7316.95	127.20	(7078.77, 7577.78)
γ_2	7136.38	70.27	(6999.10, 7277.32)
ϵ_1	0.4024	0.0126	(0.3779, 0.4273)
ϵ_2	0.3958	0.0054	(0.3852, 0.4063)
$\Delta\gamma$	180.57	144.97	(−100.15, 468.23)
$\Delta\epsilon$	0.0068	0.0136	(−0.0198, 0.0337)
$ICER$	−50599.2	5996410	(−223714.0, 191286.0)

From 20000 MCMC simulations we plot the cost–effectiveness acceptability plane (Figure 1.5). The posterior probability that T_1 increases both costs and effectiveness (quadrant I) is estimated to be 60.9%. The probabilities of the quadrants II, III and IV are 28.5%, 3.5% and 7.1%, respectively.

Although it is still a minority, the number of studies proposing the Bayesian approach in cost–effectiveness analysis is rapidly growing (Heitjan, 1997; Briggs, 1999; Heitjan et al., 1999; Al and Van Hout, 2000; O'Hagan and Stevens, 2001; O'Hagan et al., 2001; O'Hagan and Stevens, 2003; Vázquez–Polo et al., 2004, 2005b; Moreno et al., 2010, 2012, 2013b, 2014, 2016; Baio, 2014; Baio et al., 2017, among others).

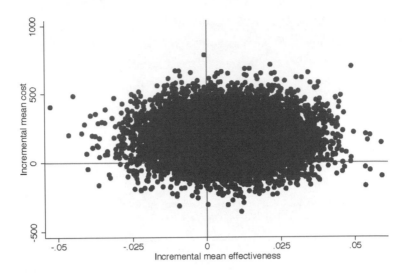

FIGURE 1.5

Cost–effectiveness plane. Bayesian analysis in Example 1.1.

Cooper et al. (2013) examined the use of implicit and explicit Bayesian methods in HTA published by the UK National Institute of Health Research between 1997 and 2011. They conclude that 41% of HTA reports contained a Bayesian analysis and 17% of them are explicit. The percentage increased from 0% in 1997 to 80% in 2011.

A simple search in MEDLINE.[2] shows the growing interest in the scientific literature concerning the use of Bayesian methods in cost–effectiveness analysis (Figure 1.6).

---

[2]Search strategy: ((bayesian) AND cost-effectiveness[Title/Abstract]) OR ((bayes) AND cost-effectiveness[Title/Abstract]) OR ((bayesian) AND cost-utility[Title/Abstract]) OR ((bayes) AND cost-utility[Title/Abstract]) OR ((bayesian) AND cost-benefit[Title/Abstract]) OR ((bayes) AND cost-benefit[Title/Abstract]) OR ((bayesian) AND economic evaluation[Title/Abstract]) OR ((bayes) AND economic evaluation[Title/Abstract]) (https://www.ncbi.nlm.nih.gov/pubmed/)

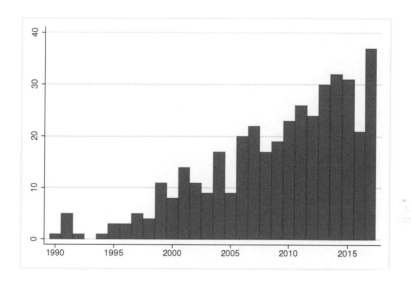

FIGURE 1.6

Number of references in MEDLINE to Bayesian analysis in cost–effectiveness analysis from 1990 to 2017.

2

Statistical inference in parametric models

2.1 Introduction

This chapter deals with parametric models, that is, with parametric distributions of an observable random variable X. The basic notions in parametric statistical inference, such as the random sample from a model, the likelihood function of the parameters of the model, and the maximum likelihood estimator of the parameters, are introduced. Sometimes the interest is in a subparameter, that is, a function of the parameters of the model. The notion of the likelihood of a subparameter is revised and the convention needed for defining the maximum likelihood estimator of the subparameter is given. The introduction of a subparameter is also known as a reparametrization of the original sampling model.

If the parametric sampling model is seen as the distribution of the random variable X, conditional on a unknown parameter θ, and if we add a prior distribution for θ, a parametric Bayesian model arises as the joint distribution of the observable variable X and the nonobservable parameter θ. In this setting, methods for eliciting prior distributions, Bayesian parameter estimation, hypotheses testing, and prediction will be discussed.

We remark that the predictive distribution of the random variable X plays a central role in statistical decision theory, and, in particular, in cost–effectiveness analysis, and hence we present both the frequentist and the Bayesian predictive distribution of X, conditional on a sample from the true model of X.

Model selection appears in a natural way in most of the statistical applications to real problems. For instance, in hypotheses testing problems more than one parametric model for the random variable X are involved and the problem is that of choosing one of them. We note that for estimating parameters from samples of moderate or large sample sizes, the frequentist and Bayesian approaches typically provide estimators close to each other. However, for hypotheses testing, the frequentist and Bayesian approaches give us results that substantially differ no matter what the sample size is. Bayesian model selection extends the hypotheses testing methodology and is clearly preferable to the frequentist approach as it is able to account for the sample size and the dimension of the parameter space. Further, the Bayesian procedure gives us an easy meaningful measure of the uncertainty in the selection of a model. As a consequence, in this chapter we only consider the Bayesian approach to model selection.

A first reason to include a background on model selection in this book is that clustering samples is a model selection problem and, at the same time, it is an important statistical problem in cost–effectiveness analysis for heterogeneous data. When the samples of cost and effectiveness of treatments come from patients from different health-care centers, they are an aggregate of samples that are typically heterogeneous, a frequent scenario in cost–effectiveness analysis. This topic will be considered later in Chapter 5.

A second reason to include model selection in the book is that the variable selection problem is a model selection problem, and, at the same time, it is a central problem in linear models that are the main tool in cost–effectiveness analysis for subgroups. The variable selection problem consists of selecting a subset of influential covariates based on the information provided by a sample of the response variable and the covariates associated with the response. This problem will be considered later in Chapter 6, where the subgroup analysis is developed. In this chapter we only outline the interest of this problem in cost–effectiveness analysis.

2.2 Parametric sampling models

Let X be an observable random variable that takes values in the sample space $\mathcal{X}$ which is either $\mathbb{R}^p$ or a subset of $\mathbb{R}^p$ for some $p \geq 1$, where $\mathbb{R}$ denotes the real line. In the context of cost–effectiveness analysis of treatments, the random variable X is a 2–dimensional vector (c, e), where c represents the cost and e the effectiveness of the treatment.

The full description of a random variable X is provided by its probability distribution, which can be represented by either a probability density $f(x)$ if X is a continuous variable, or by a probability function $p(x)$ if X is discrete, in which case $\mathcal{X}$ is a countable set. The probability density satisfies

$$f(x) \geq 0, \text{ for any } x \in \mathcal{X}, \tag{2.1}$$

and

$$\int_{\mathcal{X}} f(x)\, dx = 1. \tag{2.2}$$

For the discrete case, the probability function satisfies

$$p(x) > 0, \text{ for any } x \in \mathcal{X}, \tag{2.3}$$

and

$$\sum_{x \in \mathcal{X}} p(x) = 1. \tag{2.4}$$

The distribution of X is called the *sampling model*.

In statistical applications, the probability distribution of X is empirically chosen, and typically depends on a nonobservable parameter θ that belongs to a parameter space Θ, which is either $\mathbb{R}^p$ or an open set in $\mathbb{R}^p$ for some $p \geq 1$. This unknown parameter θ should have, however, a physical meaning. Thus, a statistical application for a continuous random variable X starts with a parametric class of sampling densities $\mathcal{F} = \{\, f(x|\theta),\ \theta \in \Theta \,\}$ that describes the random variable X up to the unknown parameter $\theta \in \Theta$. These densities must satisfy the conditions:

$$f(x|\theta) \geq 0, \text{ for any } (x, \theta) \in \mathcal{X} \times \Theta, \tag{2.5}$$

and

$$\int_{\mathcal{X}} f(x|\theta) \, dx = 1, \text{ for any } \theta \in \Theta. \tag{2.6}$$

The probability that X is valued in a Borel set $A \subset \mathbb{R}^p$, or equivalently the probability of the event A, is given by

$$\Pr(A|\theta) = \int_A f(x|\theta) \, dx,$$

which depends on the unknown θ.

If the random variable X is discrete, it is valued in a countable sample space $\mathcal{X} = \{1, 2, \dots\}$, and the sampling distribution for X is a family of probability functions $\mathcal{P} = \{\, p(x|\theta), \ \theta \in \Theta \,\}$ satisfying $0 < p(x|\theta) \leq 1$ for any $(x, \theta) \in \mathcal{X} \times \Theta$, and $\sum_{x \in \mathcal{X}} p(x|\theta) = 1$ for any $\theta \in \Theta$. In this case the probability that X is valued in a set $A \subset \mathcal{X}$ is given by

$$\Pr(A|\theta) = \sum_{x \in \mathcal{A}} p(x|\theta).$$

2.2.1   The likelihood function

To accommodate a sampling model $f(x|\theta) \in \mathcal{F}$ to a specific data set $\mathbf{x} = (x_1, \dots, x_n)$ of size n, the unknown parameter θ needs to be estimated using that data. It is assumed that the data have been independently drawn from the sampling distribution $f(x|\theta)$, where θ is the unknown true value. The sample $\mathbf{x}$ is called a random sample from $f(x|\theta)$ and because of the independence assumption, the probability density of $\mathbf{x}$ is given by

$$f(\mathbf{x}|\theta) = \prod_{i=1}^{n} f(x_i|\theta). \tag{2.7}$$

When X is discrete, the random sample $\mathbf{x}$ is drawn from a probability distribution $p(x|\theta)$ in $\mathcal{P} = \{\, p(x|\theta), \ \theta \in \Theta \,\}$, and because of the independent assumption, the joint probability of $\mathbf{x}$ is

$$p(\mathbf{x}|\theta) = \prod_{i=1}^{n} p(x_i|\theta). \tag{2.8}$$

A central notion in statistical inference is the likelihood of θ for the random sample $\mathbf{x}$, which was introduced by Fisher (1922).

Definition 2.1. *Let* $\mathbf{x} = (x_1, \ldots, x_n)$ *be a random sample from a model in the class* $\mathcal{F} = \{ f(x|\theta), \ \theta \in \Theta \}$. *Then, the likelihood of* θ *for the sample* $\mathbf{x}$ *is defined as*

$$\ell_{\mathbf{x}}(\theta) = k \, f(\mathbf{x}|\theta) = k \prod_{i=1}^{n} f(x_i|\theta), \qquad \theta \in \Theta, \tag{2.9}$$

where k *is an arbitrary positive constant. We note that the argument of the likelihood function is* θ.

When X *is discrete, the likelihood of* θ *becomes*

$$\ell_{\mathbf{x}}(\theta) = k \, p(\mathbf{x}|\theta) = k \prod_{i=1}^{n} p(x_i|\theta). \tag{2.10}$$

Since the likelihood of θ is a statistical tool for comparing different values of θ, the arbitrary constant k is not relevant.

We remark that the term likelihood is not equivalent to the term probability, although when X is a discrete random variable, the likelihood of θ for the sample $\mathbf{x}$ means the probability of the sample $\mathbf{x}$ conditional on θ.

The likelihood of θ depends on the data $\mathbf{x}$, although different data sets with the same sample size n can give the same likelihood, and hence the likelihood does not necessarily change as $\mathbf{x}$ changes.

We illustrate the likelihood function on some distributions commonly used in cost–effectiveness analysis.

The effectiveness of a treatment is sometimes described by a dichotomous random variable that indicates that a medical treatment is successful or unsuccessful when applied to a patient. An appropriate sampling distribution for that variable is the Bernoulli distribution.

Example 2.1 (Bernoulli sampling model). *Let* X *be a discrete random variable.* X *follows the Bernoulli distribution,* $Be(x|\theta)$, *if its distribution is given by*

$$Be(x|\theta) = \theta^x (1 - \theta)^{1-x}, \quad x \in \{0, 1\}, \tag{2.11}$$

where the parameter θ is in the space $\Theta = (0,1)$. The meaning of θ is the probability of success, that is, the probability that $X = 1$.

The mean and variance of X are given by $\mathbb{E}(X|\theta) = \sum_{x=0}^{1} x\,Be(x|\theta) = \theta$, and $\mathbb{V}(X|\theta) = \sum_{x=0}^{1}(x - \theta)^2 Be(x|\theta) = \theta(1 - \theta)$.

Let $\mathbf{x} = (x_1, \ldots, x_n)$ be a random sample from $Be(x|\theta)$. Then, the likelihood of θ is

$$\ell_{\mathbf{x}}(\theta) = \theta^{n\bar{x}}(1 - \theta)^{n - n\bar{x}}, \tag{2.12}$$

where $\bar{x} = \sum_{i=1}^{n} x_i/n$ is the sample mean. Thus, the likelihood function is a polynomial in θ that depends on $\mathbf{x}$ through $\bar{x}$, so all samples $\mathbf{x}$ of size n having the same sample mean $\bar{x}$ provide the same likelihood of θ. The mean $\bar{x}$ is called a "sufficient statistic," and the pair $(\bar{x}, n)$ contains all the sample information we need to define the likelihood. The sample size n is an "ancillary statistic." Ancillary statistic means that its distribution does not depend on the parameter θ.

A plot of the likelihood function of θ for any sample $\mathbf{x}$ such that $(\bar{x}, n) = (0.2, 10)$ is given in Figure 2.1. From this curve it follows that the likelihood is a bounded, unimodal function and its absolute maximum is attained at point $\theta = 0.2$.

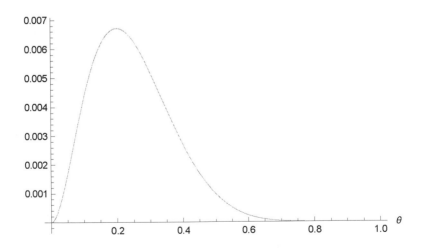

FIGURE 2.1

Likelihood of θ for $\bar{x} = 0.2$ and $n = 10$ in Example 2.1.

The effectiveness of a treatment is sometimes measured by a discrete positive variable, for instance, the number of hospitalization days of the patient receiving the treatment, or survival days, etc. A Poisson distribution might then be an appropriate distribution to model this discrete variable.

Example 2.2 (Poisson sampling model). *Let X be a discrete random variable. X follows the Poisson distribution, $Po(x|\lambda)$, if its distribution is given by*

$$Po(x|\lambda) = \frac{\lambda^x}{x!} \exp(-\lambda), \quad x = 0, 1, 2, \ldots \qquad (2.13)$$

where the parameter λ is in the space $\mathbb{R}^+$. The mean and variance of X are $\mathbb{E}(X|\lambda) = \lambda$ and $\mathbb{V}(X|\lambda) = \lambda$.

Let $\mathbf{x} = (x_1, \ldots, x_n)$ be a random sample from $Po(x|\lambda)$. Then, the likelihood of λ is

$$\ell_{\mathbf{x}}(\lambda) = \lambda^{n\bar{x}} \exp(-n\lambda), \qquad (2.14)$$

where $\bar{x} = \sum_{i=1}^{n} x_i/n$ is the sample mean. Again, all samples $\mathbf{x}$ of size n having the same sample mean $\bar{x}$ provide the same likelihood.

A plot of the likelihood function of λ for any sample $\mathbf{x}$ such that $(\bar{x}, n) = (1.5, 10)$ is given in Figure 2.2 showing that the likelihood is a bounded, unimodal function and its absolute maximum is attained at point $\lambda = 1.5$.

The normal distribution is one of the most used distributions in cost–effectiveness analysis to model either the cost or the effectiveness of a treatment.

Example 2.3 (Normal sampling model). *A continuous random variable X is normally distributed with parameters μ and σ, $\mathcal{N}(x|\mu, \sigma^2)$, if its probability density function is given by*

$$\mathcal{N}(x|\mu, \sigma^2) = \frac{1}{\sigma\sqrt{2\pi}} \exp\left\{-\frac{(x-\mu)^2}{2\sigma^2}\right\}, \quad x \in \mathbb{R}. \qquad (2.15)$$

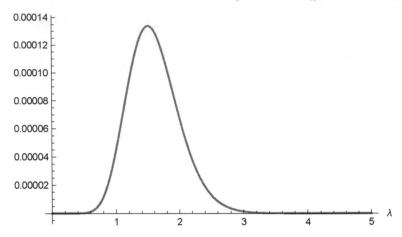

FIGURE 2.2

Likelihood of λ for $\bar{x} = 1.5$ and $n = 10$ in Example 2.2.

The parameter (μ, σ) is in the space $\Theta = \mathbb{R} \times \mathbb{R}^{+}$. The mean and variance of X are $\mathbb{E}(X|\mu, \sigma^2) = \mu$ and $\mathbb{V}(X|\mu, \sigma^2) = \sigma^2$.

Let $\mathbf{x} = (x_1, \ldots, x_n)$ be a random sample from the distribution $\mathcal{N}(x|\mu, \sigma^2)$. The likelihood of (μ, σ) is given by

$$\ell_{\mathbf{x}}(\mu, \sigma) = \sigma^{-n} \exp\left\{ -\frac{ns^2}{2\sigma^2} \right\} \exp\left\{ -\frac{n(\bar{x} - \mu)^2}{2\sigma^2} \right\}, \qquad (2.16)$$

where $\bar{x} = \sum_{i=1}^{n} x_i/n$ and $ns^2 = \sum_{i=1}^{n}(x_i - \bar{x})^2$. Thus, all samples $\mathbf{x}$ having the same $(\bar{x}, s, n)$ provide the same likelihood.

A plot of the likelihood function of (μ, σ) for any sample $\mathbf{x}$ such that $(\bar{x}, s, n) = (2, 2, 10)$ is given in Figure 2.3. The likelihood is again a bounded, unimodal function and its absolute maximum is attained at point $(\mu, \sigma) = (2, 2)$.

The lognormal distribution is typically used to model the cost of a treatment.

Example 2.4 (Lognormal sampling model). *A positive continuous random variable X is lognormally distributed, $\Lambda(x|\mu, \sigma^2)$, if its probability density function is given by*

$$\Lambda(x|\mu, \sigma) = \frac{1}{x\sigma\sqrt{2\pi}} \exp\left\{ -\frac{1}{2}\frac{(\log(x) - \mu)^2}{\sigma^2} \right\}, \quad x > 0. \qquad (2.17)$$

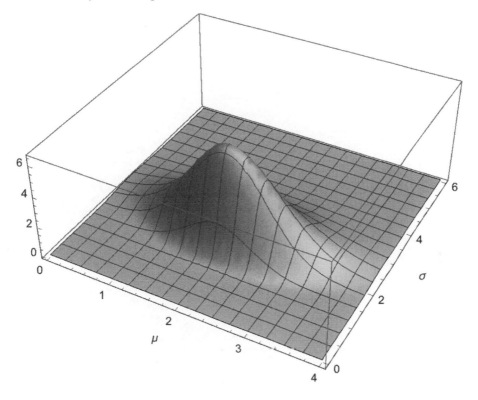

FIGURE 2.3

The likelihood of (μ, σ) for $(\bar{x}, s, n) = (2, 2, 10)$ in Example 2.3.

The parameter (μ, σ) is in the space $\Theta = \mathbb{R} \times \mathbb{R}^+$. The mean and variance of X are $\mathbb{E}(X|\mu, \sigma^2) = \exp(\mu + \sigma^2/2)$ and $\mathbb{V}(X|\mu, \sigma^2) = (\exp(\sigma^2) - 1) \exp(2\mu + \sigma^2)$.

For a sample $\mathbf{x} = (x_1, \ldots, x_n)$ the likelihood of (μ, σ) is

$$\ell_{\mathbf{x}}(\mu, \sigma) = \sigma^{-n} \exp\left\{-\frac{nv^2}{2\sigma^2}\right\} \exp\left\{-\frac{n(\mu - m)^2}{2\sigma^2}\right\}, \qquad (2.18)$$

where $m = \sum_{i=1}^n \log(x_i)/n$ and $nv^2 = \sum_{i=1}^n (\log(x_i) - m)^2$.

A plot of the likelihood function of (μ, σ) for any sample $\mathbf{x}$ such that $(m, v, n) = (2, 1, 10)$ is given in Figure 2.4. The likelihood is a bounded, unimodal function and its absolute maximum is attained at point $(\mu, \sigma) = (2, 1)$.

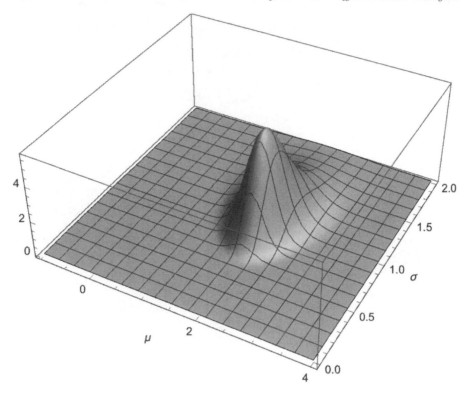

FIGURE 2.4

The likelihood of (μ, σ) for $(m, v, n) = (2, 1, 10)$ in Example 2.4.

An alternative to the lognormal distribution for modeling the cost of a treatment is the gamma distribution.

Example 2.5 (Gamma sampling model). *Let X be a positive continuous random variable. X is gamma distributed, $\mathcal{G}(x|\alpha, \beta)$, if its probability distribution is*

$$\mathcal{G}(x|\alpha, \beta) = \frac{\beta^\alpha}{\Gamma(\alpha)} x^{\alpha-1} \exp(-\beta x), \quad x > 0, \qquad (2.19)$$

where α is a shape parameter, β a scale parameter, (α, β) are in the space $\Theta = \mathbb{R}^+ \times \mathbb{R}^+$, and

$$\Gamma(z) = \int_0^\infty t^{z-1} \exp(-t) \, dt$$

is the gamma function. The mean and variance of X are $\mathbb{E}(X|\alpha, \beta) = \alpha/\beta$ and $\mathbb{V}(X|\alpha, \beta) = \alpha/\beta^2$.

Let $\mathbf{x} = (x_1, \ldots, x_n)$ be a sample from $\mathcal{G}(x|\alpha, \beta)$ in (2.19), then the likelihood function of (α, β) is

$$\ell_{\mathbf{x}}(\alpha, \beta) = \frac{\beta^{n\alpha}}{\Gamma(\alpha)^n} \exp\left(n(\alpha - 1)m\right) \exp\left(-n\beta\bar{x}\right), \qquad (2.20)$$

where $\bar{x} = \sum_{i=1}^{n} x_i/n$ and $m = \sum_{i=1}^{n} \log(x_i)/n$. Thus, all samples $\mathbf{x}$ having the same $(\bar{x}, m, n)$ provide the same likelihood. A plot of the likelihood function of (α, β) for any sample $\mathbf{x}$ such that $(\bar{x}, m, n) = (1, 0, 10)$ is given in Figure 2.5. Again, we can observe that the likelihood is a bounded, unimodal function and its absolute maximum is attained at point $(\alpha, \beta) = (1, 1)$.

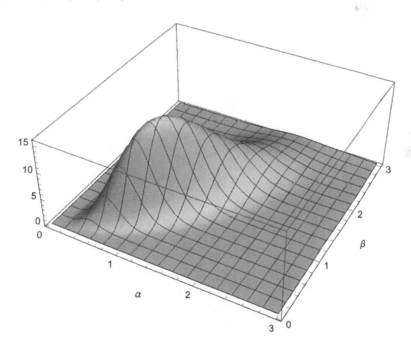

FIGURE 2.5

The likelihood of (α, β) for $(\bar{x}, m, n) = (1, 0, 10)$ in Example 2.5.

The likelihood function for most of the usual models attains its absolute maximum at a unique point as it occurs in the previous models.

We have to go to a peculiar model to find a likelihood function that attains its absolute maximum at several different points. Here is an example of that model.

Example 2.6. *Let X be a continuous random variable that follows the mixture of the Cauchy distributions $\mathcal{C}(x|\mu_1, 1)$ and $\mathcal{C}(x|\mu_2, 1)$,*

$$f(x|\mu_1, \mu_2) = 0.2 \frac{1}{\pi \left(1 + (x - \mu_1)^2\right)} + 0.8 \frac{1}{\pi \left(1 + (x - \mu_2)^2\right)}, \ x \in \mathbb{R}.$$

The parameter (μ_1, μ_2) is in the space $\Theta = \mathbb{R}^2$.

For the random sample $\mathbf{x} = (x_1, \ldots, x_n)$ from the distribution $f(x|\mu_1, \mu_2)$, the likelihood of (μ_1, μ_2) is

$$\ell_{\mathbf{x}}(\mu_1, \mu_2) = \prod_{i=1}^{n} \left(\frac{0.2}{(1 + (x_i - \mu_1)^2)} + \frac{0.8}{(1 + (x_i - \mu_2)^2)} \right).$$

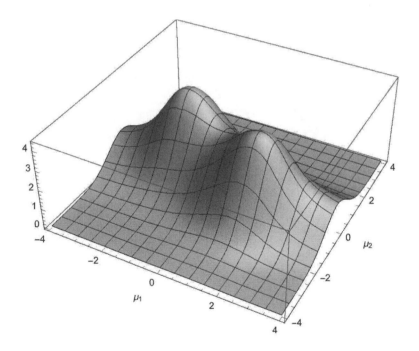

FIGURE 2.6

The likelihood of (μ_1, μ_2) for $\mathbf{x} = (-2, -1, 0, 1, 2)$ in Example 2.6.

A plot of this likelihood function for the data $\mathbf{x} = (-2, -1, 0, 1, 2)$ *in the region* $(\mu_1, \mu_2) \in (-4, 4) \times (-4, 4)$ *is given in Figure 2.6, and it shows that its absolute maximum is attained at points* $(\mu_1, \mu_2) = (-1.53, 0.57)$ *and* $(\mu_1, \mu_2) = (1.53, -0.57)$.

2.2.2 Likelihood sets

The likelihood function $\ell_{\mathbf{x}}(\theta)$ of a parameter θ for a given sample $\mathbf{x}$ provides an order of preference among different values of θ. Those θ having a large likelihood are preferred as candidates to estimate the value of θ from which the data $\mathbf{x}$ come. This assertion is but a consequence of Lemma 2.1 in Section 2.2.3. When X is discrete, a point θ with a large likelihood assigns a large probability to the observed sample $\mathbf{x}$.

A distinguished subset of Θ is the so–called *likelihood set* $K(g, \mathbf{x})$ for a level $g > 0$ and sample $\mathbf{x}$. The set $K(g, \mathbf{x})$ is defined by the condition that for any point θ outside of $K(g, \mathbf{x})$ there is one point θ' inside of $K(g, \mathbf{x})$ having a larger likelihood. Thus, $K(g, \mathbf{x})$ is defined by

$$K(g, \mathbf{x}) = \left\{ \theta \; : \; f(\mathbf{x}|\theta) \geq g \right\}. \tag{2.21}$$

It is clear that the wider the likelihood set $K(g, \mathbf{x})$ is, the flatter the likelihood function $f(\mathbf{x}|\theta)$, and hence the smaller the discrimination between values of θ.

Example 2.7 (Lognormal likelihood sets). *Let us suppose we have the sample* $\mathbf{x} = (2.17, 0.99, 1.29, 7.05, 2.33)$ *from a lognormal sampling model with unknown parameters* μ *and* σ, *for which* $(m, v^2, n) = (0.76, 0.46, 5)$. *Figure 2.7 shows the likelihood sets for several values of* g. *In the middle of the figure, the darkest grey likelihood set corresponds to level* $g = 0.00012$.

Interesting Bayesian properties of the likelihood sets are presented in Section 2.3.

A distinguished point in the likelihood sets is the one that maximized the likelihood function that we present in the next subsection.

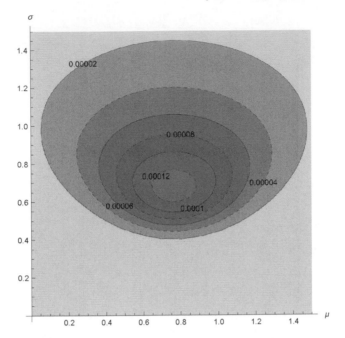

FIGURE 2.7

Likelihood sets of (μ, σ) for $\mathbf{x} = (2.17, 0.99, 1.29, 7.05, 2.33)$ and levels $g = \{2\ 10^{-5}, 4\ 10^{-5}, 6\ 10^{-5}, 8\ 10^{-5}, 10^{-4}, 1.2\ 10^{-4}\}$ in Example 2.7.

2.2.3   The maximum likelihood estimator

Let $\mathbf{x} = (x_1, \ldots, x_n)$ be a random sample from a model in $\mathcal{F} = \{f(x|\theta),\ \theta \in \Theta\}$. It is assumed in what follows that the likelihood function $\ell_{\mathbf{x}}(\theta)$ reaches its absolute maximum at just one point.

Definition 2.2. *The maximum likelihood estimator (MLE) of θ for the sample $\mathbf{x}$ is the point $\hat{\theta}(\mathbf{x})$ that maximizes the likelihood function, that is*

$$\hat{\theta}(\mathbf{x}) = \arg \sup_{\theta \in \Theta} \ell_{\mathbf{x}}(\theta), \tag{2.22}$$

assuming that this point is unique.

A quite simple argument suggests that this estimator typically gives an accurate estimation of the true value of θ for large sample sizes *n*. The proof follows from the general Kullback–Leibler information inequality that we prove in the next lemma.

Lemma 2.1. *If $f(x)$ and $g(x)$ are two probability densities, the following inequality*

$$\mathbb{E}_{f(x)}\left(-\log\left(\frac{g(x)}{f(x)}\right)\right) = \int_{\mathcal{X}} f(x) \log\left(\frac{f(x)}{g(x)}\right) \, dx \geq 0$$

holds. The inequality is strict except for $f(x) = g(x)$.

Proof. Since $-\log(x)$ is a convex function, it follows from Jensen's inequality that

$$\mathbb{E}_{f(x)}\left(-\log\left(\frac{g(x)}{f(x)}\right)\right) \geq -\log\left(\mathbb{E}_{f(x)}\left(\frac{g(x)}{f(x)}\right)\right)$$
$$= -\log\left(\int_{\mathcal{X}} g(x)\, dx\right) = 0,$$

and this proves the assertion. ☐

If we write the information inequality for $f(x) = f(x|\theta_0)$ and $g(x) = f(x|\theta)$ we have that

$$\mathbb{E}_{x|\theta_0}\left(\log(f(x|\theta_0))\right) \geq \mathbb{E}_{\mathbf{x}|\theta_0}\left(\log(f(x|\theta))\right).$$

Then, applying the Law of Large Numbers, it follows that when sampling from $f(x|\theta_0)$ and the sample size n is large, the sample means of $\{\log(f(x_i|\theta_0)), i = 1, ..., , n\}$ and $\{\log(f(x_i|\theta)), i = 1, ..., n\}$ satisfy the inequality

$$\frac{1}{n}\sum_{i=1}^{n}\log(f(x_i|\theta_0)) \geq \frac{1}{n}\sum_{i=1}^{n}\log(f(x_i|\theta)), \qquad [P_{\theta_0}].$$

The notation $[P_{\theta_0}]$ means that when the samples come from the distribution P_{θ_0}, the probability that this inequality holds computed with the distribution P_{θ_0} tends to 1 as n grows. Therefore, for large n we have

$$\prod_{i=1}^{n} f(x_i|\theta_0) \geq \prod_{i=1}^{n} f(x_i|\theta), \qquad [P_{\theta_0}].$$

This means that for large sample size n, the true value θ_0 has a larger likelihood than that of any $\theta \neq \theta_0$. This suggests that by maximizing the likelihood function, an accurate estimator of the true θ can be obtained, which is an idea that we formalize later on.

In most practical situations, especially when the parameter is continuous, it is convenient to use the logarithm of the likelihood function. Since log is a monotone function, to find the maximum in (2.22) is equivalent to finding the maximum of the logarithm of the likelihood function. Let us illustrate the procedure on the Bernoulli sampling model.

Example 2.1 (continued). *Let* $\mathbf{x} = (x_1, \ldots, x_n)$ *be a random sample from* $Be(x|\theta)$ *in (2.11) with the likelihood function given in (2.12). Then*

$$\log(\ell_{\mathbf{x}}(\theta)) = n\bar{x}\log(\theta) + (n - n\bar{x})\log(1 - \theta),$$

where $n\bar{x} = \sum_{i=1}^{n} x_i$. *Taking the first derivative of the log–likelihood we obtain*

$$\frac{\partial \log(\ell_{\mathbf{x}}(\theta))}{\partial \theta} = \frac{n\bar{x}}{\theta} - \frac{n - n\bar{x}}{1 - \theta} = 0.$$

The unique root of this equation is $\hat{\theta} = \bar{x}$, *and it turns out to be the absolute maximum of the likelihood function.*

To maximize the likelihood function is an optimization problem that typically presents some difficulties. For instance, in some situations the log–likelihood function is not differentiable everywhere in Θ or the equation may have more than one root. In other situations, the equation has to be solved numerically. Some of these difficulties are illustrated using the gamma model.

Example 2.5 (continued). *Let* $\mathbf{x} = (x_1, \ldots, x_n)$ *be a sample from* $\mathcal{G}(x|\alpha, \beta)$ *where* α, β *are unknown. The* log *of the likelihood function is given by*

$$\log(\ell_{\mathbf{x}}(\alpha, \beta)) = n\alpha\log(\beta) - n\log(\Gamma(\alpha)) + n(\alpha - 1)m - n\beta\bar{x}.$$

The first derivative yields the nonlinear equations in α and β

$$n \log(\beta) - n\frac{\Gamma'(\alpha)}{\Gamma(\alpha)} + nm = 0$$

$$\frac{n\alpha}{\beta} - n\bar{x} = 0.$$

From the second equation $\beta = \alpha/\bar{x}$, and replacing it in the first equation we have

$$\log(\alpha) - \frac{\Gamma'(\alpha)}{\Gamma(\alpha)} = \log(\bar{x}) - m. \tag{2.23}$$

The α root of this equation, $\hat{\alpha}$, cannot be obtained in close form although it can be numerically obtained. It can be shown that $\hat{\alpha}$ and $\hat{\beta} = \hat{\alpha}/\bar{x}$ are the MLEs of α and β.

Table 2.1 displays the MLEs for the parameters of some commonly used sampling models in cost–effectiveness analysis.

To be able to show that the maximum likelihood estimators of the parameter θ of the family of densities $\mathcal{F}$ have certain desirable properties for large samples sizes, the densities are required to be *regular*, that is, Conditions 1 to 5 below have to be satisfied. For simplicity we assume for the moment that Θ is either $\mathbb{R}$ or an open set in $\mathbb{R}$.

Condition 1. *The sample space $\mathcal{X}$ does not depend on θ.*

Condition 2. *The equalities*

$$\frac{d}{d\theta} \int_{\mathcal{X}} f(x|\theta) \, dx = \int_{\mathcal{X}} \frac{d}{d\theta} f(x|\theta) \, dx, \tag{2.24}$$

and

$$\frac{d^2}{d\theta^2} \int_{\mathcal{X}} f(x|\theta) \, dx = \int_{\mathcal{X}} \frac{d^2}{d\theta^2} f(x|\theta) \, dx, \tag{2.25}$$

hold for any $\theta \in \Theta$.

Condition 3. *The function*

$$I(\theta) = \int_{\mathcal{X}} \left(-\frac{d^2 \log(f(x|\theta))}{d\theta^2} \right) f(x|\theta) \, dx, \tag{2.26}$$

is positive for any $\theta \in \Theta$.

TABLE 2.1

MLE of the parameters of some useful sampling models in cost–effectiveness analysis.

Sampling model	Parameter	MLE
$\mathrm{Be}(x\|\theta)$	θ	$\hat{\theta} = \bar{x}$
$\mathrm{Po}(x\|\lambda)$	λ	$\hat{\lambda} = \bar{x}$
$\mathcal{N}(x\|\mu, \sigma_0^2)$, σ_0^2 known	μ	$\hat{\mu} = \bar{x}$
$\mathcal{N}(x\|\mu_0, \sigma^2)$, μ_0 known	σ^2	$\hat{\sigma}^2 = \dfrac{1}{n}\sum_{i=1}^{n}(x_i - \mu_0)^2$
$\mathcal{N}(x\|\mu, \sigma^2)$	μ, σ^2	$\hat{\mu} = \bar{x}$, $\hat{\sigma}^2 = \dfrac{1}{n}\sum_{i=1}^{n}(x_i - \bar{x})^2$
$(y_i = \log(x_i))$		
$\Lambda(x\|\mu, \sigma_0^2)$, σ_0^2 known	μ	$\hat{\mu} = \bar{y}$
$\Lambda(x\|\mu_0, \sigma^2)$, μ_0 known	σ^2	$\hat{\sigma}^2 = \dfrac{1}{n}\sum_{i=1}^{n}(y_i - \mu_0)^2$
$\Lambda(x\|\mu, \sigma^2)$	μ, σ^2	$\hat{\mu} = \bar{y}$, $\hat{\sigma}^2 = \dfrac{1}{n}\sum_{i=1}^{n}(y_i - \bar{y})^2$
$\mathcal{G}(x\|\alpha, \beta_0)$, β_0 known	α	$\hat{\alpha}$ such that $\dfrac{\Gamma'(\hat{\alpha})}{\Gamma(\hat{\alpha})} = \log(\beta_0) + \bar{y}$
$\mathcal{G}(x\|\alpha_0, \beta)$, α_0 known	β	$\hat{\beta} = \dfrac{\alpha_0}{\bar{x}}$
$\mathcal{G}(x\|\alpha, \beta)$	α, β	$\hat{\alpha}$ in (2.23) $\hat{\beta} = \dfrac{\hat{\alpha}}{\bar{x}}$

Condition 4. *The random function*

$$S(x|\theta) = \frac{d}{d\theta}\log(f(x|\theta)) \qquad (2.27)$$

is a continuous function of $\theta \in \Theta$.

Condition 5. *For any $\theta_0 \in \Theta$, the function*

$$A(\theta, \theta_0) = \int_{\mathcal{X}} S(x|\theta) \, f(x|\theta_0) \, dx, \qquad (2.28)$$

is a continuous function of θ.

We note that Condition 1 excludes, for instance, the uniform sampling density

$$f(x|\theta) = \frac{1}{\theta} \, \mathbf{1}_{(0,\theta)}(x), \qquad (2.29)$$

where $\theta \in \mathbb{R}^+$. However, the likelihood of θ for the sample $\mathbf{x}$ attains its absolute maximum at point $\hat{\theta}(\mathbf{x}) = \max_{i=1,\ldots,n} x_i$, which is a quite stable estimator. Nevertheless, when Condition 1 is not satisfied, the MLE can be very unstable.

For instance, Condition 1 fails in the binomial distribution with n and θ unknown. We note that $n \in \{x, x+1, \ldots\}$. Let us illustrate this instability on a binomial example taken from Olkin et al. (1981).

Example 2.8 (Binomial sampling model with two unknown parameters). *Let X be a random variable with binomial distribution*

$$Bin(x|n, \theta) = \binom{n}{x} \theta^x (1-\theta)^{n-x}, \ 0 < \theta < 1, \ x = 0, 1, \ldots, n, \quad (2.30)$$

where n and θ are unknown parameters.

Let $\mathbf{x} = (x_1, \ldots, x_N)$ be a random sample from $Bin(x|n, \theta)$. The MLE $\hat{n}$ and $\hat{\theta}$ are the solutions to the nonlinear equations (Johnson et al., 2005, p. 129)

$$\hat{n} \, \hat{\theta} = \bar{x},$$

$$\sum_{j=0}^{\max_{i=1,\ldots,N}(x_i)} \frac{A_j}{\hat{n} - j} = -N \log \left(1 - \frac{\bar{x}}{\hat{n}} \right),$$

where A_j is the number of observations that exceed j.

For the data set $\mathbf{x}_1 = (16, 18, 22, 25, 27)$ the MLE of n is $\hat{n}(\mathbf{x}_1) = 99$. If the data set changes to $\mathbf{x}_2 = (16, 18, 22, 25, 28)$, the MLE of n is $\hat{n}(\mathbf{x}_2) = 190$. However, the only difference between $\mathbf{x}_1$ and $\mathbf{x}_2$ is the last observation, which is 27 in $\mathbf{x}_1$ and 28 in $\mathbf{x}_2$. This shows *the large instability of the MLE for estimating n. As a consequence, instability is also present in the estimator of θ, in fact $\hat{\theta}(\mathbf{x}_1) = 0.218$ and $\hat{\theta}(\mathbf{x}_2) = 0.114$.*

We will go back on this example to compare the MLEs of n and θ with the Bayesian estimators for the samples $\mathbf{x}_1$ and $\mathbf{x}_2$.

The likelihood function $\ell_{\mathbf{x}}(\theta)$ does not necessarily reach its absolute maximum at just one point, even when the above regularity conditions hold as shown in Example 2.6. In that case the MLE is of no interest. However, most of the models we usually use in statistical applications have a unique MLE. In those cases the MLEs are "excellent" estimators of θ providing that the sample size n is relatively large. The asymptotic properties of the MLEs are summarized in the next theorem.

Theorem 2.1. *Let $(x_1, \ldots, x_n)$ be n independent and identically distributed (i.i.d.) random variables with distribution $f(x|\theta_0) \in \mathcal{F}$, where θ_0 is an arbitrary but fixed point in Θ. Then,*

a) the random MLE sequence $\{\hat{\theta}(x_1, \ldots, x_n),\ n \geq 1\}$ degenerates in probability $[P_{\theta_0}]$ to the point θ_0 as n goes to infinity, that is

$$\lim_{n\to\infty} \hat{\theta}(x_1, \ldots, x_n) = \theta_0, \quad [P_{\theta_0}],$$

b) the limiting distribution of $n^{1/2}(\hat{\theta}(x_1, \ldots, x_n) - \theta_0)$ is the normal distribution with zero mean and variance $I^{-1}(\theta_0)$.

An estimator satisfying a) is said to be *consistent*, and if it also satisfies b), is said to be *asymptotically normal*. Consistency of an estimator means that when the sample size n grows to infinity, our uncertainty on the estimate of θ disappears. An inconsistent estimator should consequently be rejected.

On the other hand, when the asymptotic normality of an estimator is present, the normal distribution is typically used as an approximation of the distribution of the estimator for moderate sample sizes. This approximation plays a central role in hypothesis testing.

The asymptotic properties of the MLE in Theorem 2.1 can be extended to the case where $\Theta = \mathbb{R}^p$ for $p > 1$. In that case $\hat{\boldsymbol{\theta}}(x_1, \ldots, x_n)$

is a vector of dimension p for any $n \geq 1$ and when sampling from $f(x|\boldsymbol{\theta_0})$, the limit in probability of the sequence $\{\hat{\boldsymbol{\theta}}(x_1, \ldots, x_n), \; n \geq 1\}$ converges to the p−vector $\boldsymbol{\theta_0}$, that is

$$\lim_{n \to \infty} \hat{\boldsymbol{\theta}}(x_1, \ldots, x_n) = \boldsymbol{\theta_0}, \qquad [P_{\boldsymbol{\theta_0}}].$$

Further, the limiting distribution of the random vector $n^{1/2}(\hat{\boldsymbol{\theta}}(x_1, \ldots, x_n) - \boldsymbol{\theta_0})$ follows a p−variate normal distribution with mean $\mathbf{0}_p$ and covariance matrix $I^{-1}(\boldsymbol{\theta_0})$, where $I^{-1}(\boldsymbol{\theta})$ is the inverse of the $p \times p$ matrix with elements

$$I_{ij}(\boldsymbol{\theta}) = \mathbb{E}_{x|\theta}\left(-\frac{\partial^2 \log(f(x|\boldsymbol{\theta}))}{\partial\theta_i\partial\theta_j}\right)$$

for $i, j = 1, \ldots, p$ (Lehmann and Casella, 1998, Section 6.3).

2.2.3.1 Proving consistency and asymptotic normality

(*This subsection can be omitted in a first reading.*)

The random function $S(x|\theta)$ in Condition 4 is called the *score function*, and since $f(x_1, \ldots, x_n|\theta) = \prod_{i=1}^{n} f(x_i|\theta)$ we have that the score function for the sample $(x_1, \ldots, x_n)$ is given by

$$S(x_1, \ldots, x_n|\theta) = \sum_{i=1}^{n} S(x_i|\theta).$$

Function $I(\theta)$ in Condition 3 is called the *Fisher information* of the sampling models $f(x|\theta)$ and represents the expectation $\mathbb{E}_{x|\theta}(-d^2 \log(f(x|\theta))/d\theta^2)$. From Condition 1 it follows that the Fisher information can also be written as $\mathbb{E}_{x|\theta}(d \log(f(x|\theta))/d\theta)^2$, that is

$$I(\theta) = \int_{\mathcal{X}} \left(\frac{d\log(f(x|\theta))}{d\theta}\right)^2 f(x|\theta)\, dx.$$

From Condition 3 it follows that the sample mean of $S(x|\theta)$ with respect to $f(x|\theta)$ is zero, that is

$$\mathbb{E}_{x|\theta}S(x|\theta) = \int_{\mathcal{X}} S(x|\theta)f(x|\theta)\, dx = 0,$$

and hence the sample variance is

$$\mathbb{E}_{x|\theta}S(x|\theta)^2 = I(\theta).$$

The function $A(\theta, \theta_0)$ in Condition 5 is the expectation of $\log(f(x|\theta))$ with respect to model $f(x|\theta_0)$, and for $\theta = \theta_0$ we have

$$A(\theta_0, \theta_0) = \mathbb{E}_{x|\theta_0} S(x|\theta_0) = 0.$$

The derivative of $A(\theta, \theta_0)$ with respect to θ is given by

$$\frac{d}{d\theta} A(\theta, \theta_0) = \int_{\mathcal{X}} \frac{d^2 \log(f(x|\theta))}{d\theta^2} f(x|\theta_0) \, dx,$$

and for $\theta = \theta_0$ we have

$$\frac{d}{d\theta} A(\theta_0, \theta_0) = -I(\theta_0) < 0.$$

This implies that $A(\theta, \theta_0)$ is a decreasing function of θ in the interval $(\theta_0 - \delta, \theta_0 + \delta)$ for some δ. Thus, $A(\theta, \theta_0) \geq 0$ for $\theta \in (\theta_0 - \delta, \theta_0)$ and $A(\theta, \theta_0) \leq 0$ for $\theta \geq \theta_0 + \delta$.

Then, when sampling from $f(x|\theta_0)$ it follows from the Law of the Large Numbers (LLN) that:
For $\theta \in (\theta_0 - \delta, \theta_0)$,

$$\lim_{n \to \infty} \frac{1}{n} S(x_1, \ldots, x_n; \theta) = A(\theta, \theta_0) > 0 \qquad [P_{\theta_0}],$$

and for $\theta \in (\theta_0, \theta_0 + \delta)$,

$$\lim_{n \to \infty} \frac{1}{n} S(x_1, \ldots, x_n; \theta) = A(\theta, \theta_0) < 0 \qquad [P_{\theta_0}].$$

Thus, when sampling from $f(x|\theta_0)$ taking a root $\hat{\theta}_n = \hat{\theta}(x_1, \ldots, x_n)$ of the equation

$$S(x_1, \ldots, x_n|\hat{\theta}_n) = \frac{d}{d\theta} \sum_{i=1}^{n} \log(f(x_i|\hat{\theta}_n)) = 0,$$

we have that the sequence $\{\hat{\theta}_n, \ n \geq 1\}$ converges to the true value θ_0. If the root is unique we have shown that the root of the equation

$$\ell_{x_1, \ldots, x_n}(\hat{\theta}_n) = 0,$$

is the MLE, and it converges in probability to the true value θ_0. This proves part a).

On the other hand, when sampling from $f(x|\theta_0)$ it follows from the Central Limit Theorem that, in probability P_{θ_0}, as n tends to infinity, the limiting distribution of the sequence

$$Z_n = \frac{1}{n^{1/2}} \sum_{i=1}^{n} S(x_i|\theta_0)$$

is the normal distribution with zero mean and variance $I(\theta_0)$, that is $\mathcal{N}(z|0, I(\theta_0))$. Since $S(x_i|\hat{\theta}_n)$ approximates $S(x_i|\theta_0)$ for large n, using the Taylor expansion of $S(x_i|\theta_0)$ at point $\hat{\theta}_n$ we have

$$\frac{1}{n^{1/2}} \sum_{i=1}^{n} S(x_i|\theta_0) = n^{1/2}(\hat{\theta}_n - \theta_0) \frac{1}{n} \sum_{i=1}^{n} \left(-\frac{d^2}{d\theta^2} f(x_i|\theta_n^*) \right) \quad [P_{\theta_0}],$$

where $\theta_n^* = \theta_n^*(x_1, \ldots, x_n)$ is a random variable such that

$$|\hat{\theta}_n - \theta_n^*| \le |\hat{\theta}_n - \theta_0|, \qquad [P_{\theta_0}].$$

Further, from the LLN we have that as n tends to infinity

$$\lim_{n \to \infty} \frac{1}{n} \sum_{i=1}^{n} \left(-\frac{d^2}{d\theta^2} f(x_i|\theta^*) \right) = I(\theta_0) \qquad [P_{\theta_0}],$$

and hence when sampling from $f(x|\theta_0)$ the sequences

$$\{Z_n\} \text{ and } \left\{ n^{1/2}(\hat{\theta}_n - \theta_0)I(\theta_0) \right\}$$

have the same limiting distribution (Loève, 1963, Law–equivalence Lemma, p. 278). Thus, the limiting distribution of $n^{1/2}(\hat{\theta}_n - \theta_0)$ is $\mathcal{N}(z|0, I^{-1}(\theta_0))$. This proves part b).

2.2.4 Reparametrization to a subparameter

We have seen that for the model $f(x|\theta)$ the likelihood of θ for the sample $\mathbf{x}$ is defined as $\ell_{\mathbf{x}}(\theta) = f(\mathbf{x}|\theta)$ for $\theta \in \Theta$. If we reparametrize the probability density to a subparameter $\lambda(\theta)$, where $\lambda(\theta)$ is a one–to–one transformation, the likelihood of $\lambda = \lambda(\theta)$ is certainly defined as $\ell_{\mathbf{x}}(\lambda) = \ell_{\mathbf{x}}(\theta)$, $\theta \in \Theta$. Further, it is clear that the MLE of λ is given by $\hat{\lambda} = \lambda(\hat{\theta})$.

However, when $\lambda(\theta)$ is not a one–to–one function, the notion of likelihood of $\lambda = \lambda(\theta)$ is not at all clear, and a convention is then necessary. In this case the coset of λ is the subset of Θ

$$G(\lambda) = \left\{\ \theta : \lambda(\theta) = \lambda\ \right\},$$

and hence the likelihood of λ involves the set of original likelihoods

$$\left\{\ \ell_{\mathbf{x}}(\theta), \theta \in G(\lambda)\ \right\}.$$

Thus, the likelihood of λ is not uniquely defined.

A proposal for choosing a likelihood in the set $\left\{\ell_{\mathbf{x}}(\theta), \theta \in G(\lambda)\right\}$ to be assigned as the likelihood of λ was given by Zehna (1966) who defined

$$\ell_{\mathbf{x}}(\lambda) = \sup_{\theta \in G(\lambda)} \ell_{\mathbf{x}}(\theta).$$

The idea behind this definition is to get that $\lambda(\hat{\theta})$ is the point at which $\sup_{\lambda} \ell_{\mathbf{x}}(\lambda)$ is attained. This behavior is called the invariant principle of the MLE, and we remark that this convention is accepted by the statistical community. Let us illustrate the invariance principle on a Bernoulli distribution taken from Rohatgi (1976).

Example 2.1 (continued). *Consider the Bernoulli sampling model $Be(x|\theta)$ in (2.11) and let $\lambda(\theta) = \theta(1 - \theta)$, which is not a one–to–one function.*

Since $\Theta = [0, 1], \lambda(\Theta) = [0, 1/4]$ and the MLE of θ based on a random sample $\mathbf{x}$ of size n is $\hat{\theta} = \bar{x}$. Thus, for each $0 \leq \lambda_0 \leq 1/4$, the coset of λ_0 is the two–point subset

$$G(\lambda_0) = \{\theta\ :\ \theta(1 - \theta) = \lambda_0\} = \{\theta_1, \theta_2\}\,,$$

where $\theta_1 = (1 + \sqrt{1 - 4\lambda_0})/2$, and $\theta_2 = (1 - \sqrt{1 - 4\lambda_0})/2$.

The likelihood of λ_0 is given by

$$\ell_{\mathbf{x}}(\lambda_0) = \max\left\{\ell_{\mathbf{x}}(\theta_1), \ell_{\mathbf{x}}(\theta_2)\right\}.$$

Hence the MLE of parameter λ is $\hat{\lambda} = \lambda(\hat{\theta}) = \bar{x}(1 - \bar{x})$.

2.3 Parametric Bayesian models

Let $f(x|\theta)$ be the sampling model of the observable random variable X, where θ is an unobservable unknown parameter in Θ. The Bayesian statistical approach assumes that the uncertainty one has on θ is described by a probability distribution $\pi(\theta)$. This distribution is called the *prior* distribution for θ. That is, the Bayesian approach assumes that both X and θ are random variables, although X can be observed and θ cannot.

Then, the parametric Bayesian model consists of a joint distribution $f(x, \theta)$ for the random variable (X, θ), although it is convenient to decompose the joint distribution $f(x, \theta)$ as

$$f(x, \theta) = f(x|\theta)\pi(\theta),$$

and hence a Bayesian model is specified by two elements, the sampling distribution $f(x|\theta)$, the distribution of the observable random variable X conditional on θ, and the prior distribution $\pi(\theta)$ for the nonobservable parameter θ.

Given a sample $\mathbf{x} = (x_1, \dots, x_n)$ from a model in the class $\{f(x|\theta), \theta \in \Theta\}$, the first step in Bayesian estimation is that of computing the probability distribution of θ conditional on the sample $\mathbf{x}$. This distribution is given by

$$\pi(\theta|\mathbf{x}) = \frac{f(\mathbf{x}|\theta)\pi(\theta)}{m(\mathbf{x})}, \quad \theta \in \Theta, \tag{2.31}$$

where

$$m(\mathbf{x}) = \int_\Theta f(\mathbf{x}|\theta)\pi(\theta) \, d\theta,$$

is the marginal distribution of the sample $\mathbf{x}$.

Expression (2.31) is the so–called *Bayes theorem*, and the conditional distribution $\pi(\theta|\mathbf{x})$ is called the *posterior* distribution of θ. Since the marginal distribution $m(\mathbf{x})$ does not depend on θ, we can write $\pi(\theta|\mathbf{x}) \propto f(\mathbf{x}|\theta)\pi(\theta)$, where $\propto$ means *proportional to*.

A Bayesian estimator of θ is a distinguished point $\tilde{\theta} = \tilde{\theta}(\mathbf{x})$ drawn from their posterior distribution. This point $\tilde{\theta}$ is typically the mean of the posterior distribution,

$$\tilde{\theta}(\mathbf{x}) = \mathbb{E}(\theta|\mathbf{x}) = \int_{\Theta} \theta\, \pi(\theta|\mathbf{x})\, d\theta, \qquad (2.32)$$

assuming that the integral is finite. Sometimes $\tilde{\theta}$ is the mode of $\pi(\theta|\mathbf{x})$,

$$\tilde{\theta}(\mathbf{x}) = \arg\sup_{\theta \in \Theta} \pi(\theta|\mathbf{x}). \qquad (2.33)$$

When Θ is the real line $\mathbb{R}$, a robust estimator of θ is the median of the posterior distribution, assuming that it is unique, i.e., $\tilde{\theta}(\mathbf{x})$ is such that

$$\int_{-\infty}^{\tilde{\theta}(\mathbf{x})} \pi(\theta|\mathbf{x})\, d\theta = \frac{1}{2}. \qquad (2.34)$$

Example 2.9 (Gamma sampling model with a known shape parameter). *Let* $\mathbf{x} = (x_1, \ldots, x_n)$ *be a random sample from a gamma sampling model with known shape parameter* α_0, *and unknown scale parameter* $\beta > 0$,

$$f(x|\beta) = \frac{\beta^{\alpha_0}}{\Gamma(\alpha_0)} x^{\alpha_0 - 1} \exp\left(-\beta x\right), \quad x > 0. \qquad (2.35)$$

Then, the likelihood of β *is*

$$\ell_{\mathbf{x}}(\beta) = \beta^{\alpha_0 n} \exp\left(-n\beta\bar{x}\right),$$

where $\bar{x} = \sum_{i=1}^{n} x_i/n$. *Let us assign to* β *the gamma prior,*

$$\pi(\beta) = \frac{\beta_1^{\alpha_1}}{\Gamma(\alpha_1)} \beta^{\alpha_1 - 1} \exp\left(-\beta_1 \beta\right), \quad \beta > 0,$$

where α_1 *and* β_1 *are hyperparameters. This is known as the "conjugate" prior for the family* $f(x|\beta)$, *a notion that we introduce later on in Section 2.3.2.*

From (2.31), the posterior density of β *is given by*

$$\pi(\beta|\mathbf{x}) = \frac{(n\bar{x} + \beta_1)^{\alpha_0 n + \alpha_1}}{\Gamma(\alpha_0 n + \alpha_1)} \beta^{\alpha_0 n + \alpha_1 - 1} \exp\left\{-\beta(n\bar{x} + \beta_1)\right\}, \quad \beta > 0,$$

$$(2.36)$$

which is a gamma distribution with shape parameter $\alpha_0 n + \alpha_1$ and scale $n\bar{x} + \beta_1$ that depends on the sample size, the sample mean, the known shape parameter, and the hyperparameters of the prior.

In Figure 2.8 we plot the likelihood of the parameter β, the prior and the posterior distribution for β for a sample with $\alpha_0 = 10, n = 5, \bar{x} = 1$, and hyperparameters $\alpha_1 = 3.5, \beta_1 = 0.5$.

Figure 2.8 indicates that the contribution of the likelihood to the posterior distribution of β is greater than the contribution of the prior. The mean, mode, and median of the posterior distribution of β turn out to be 9.76, 9.58, and 9.70, respectively. These values are very close to the MLE $\hat{\beta} = 10.04$.

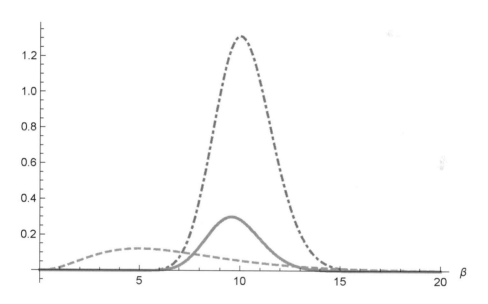

FIGURE 2.8

Triplot in Example 2.9: Likelihood function (dot-dashed), prior density (dashed), and posterior density (continuous).

We note that in Example 2.9 the posterior distribution of β, conditional on the sample $\mathbf{x}$, depends on the mean of the sample, the *sufficient statistic* of this family, and the ancillary sample size n. This is because the sufficiency notion is formulated in the Bayesian approach as follows. A statistic $T(\mathbf{x})$ is sufficient for the family of parametric distributions $\{f(x|\theta), \theta \in \Theta\}$ if and only if the equality

$$\pi(\theta|\mathbf{x}) = \pi(\theta|T(\mathbf{x}))$$

holds for any prior $\pi(\theta)$ and any sample $\mathbf{x}$.

Thus, in this book we will use indistinctly any of the two notations of the posterior distribution.

When the interest is on a subparameter $\lambda(\theta)$, the sufficiency notion is less demanding. A statistic $T(\mathbf{x})$ is sufficient for the parametric family $\{f(x|\lambda), \lambda \in \Lambda\}$ if and only if the equality

$$\pi(\lambda|\mathbf{x}) = \pi(\lambda|T(\mathbf{x}))$$

holds for any prior $\pi(\lambda)$ and any sample $\mathbf{x}$. This notion was introduced by Kolmogorov (1942) and is called K-sufficiency to distinguish it from the sufficiency notion.

While the choice of the sampling distribution $f(x|\theta)$ is accepted as an inevitable part of the statistical task, the need for a prior distribution on θ for making inference has encountered some controversies in the statistical community. The choice of the prior is perceived as the weaker chain of the Bayesian approach, and the nonobservable nature of parameter θ appears to be the responsible for that. However, arguments in favor of the Bayesian approach include i) probability is the natural tool to measure uncertainty, and this is what the prior distribution $\pi(\theta)$ does, and ii) given a sample $\mathbf{x} = (x_1, \dots, x_n)$ from a model in the class $\{f(x|\theta), \theta \in \Theta\}$, the Bayes theorem converts the prior distribution $\pi(\theta)$ into the posterior distribution $\pi(\theta|\mathbf{x})$, and by so doing sampling information is not only incorporated into the posterior distribution for θ, but also the prior belief on θ contained in $\pi(\theta)$. This probabilistic way of learning on the parameter from the data is sequential: given a posterior distribution, new data update this posterior distribution, and so on. That is, for a given pair of independent data x_1 and x_2, conditional on θ, and drawn from $f(x|\theta)$, we have that

$$\pi(\theta|x_1, x_2) = \frac{f(x_1, x_2|\theta)\pi(\theta)}{m(x_1, x_2)} = \frac{f(x_1|\theta)\pi(\theta)}{m(x_1)} m(x_1) \frac{f(x_2|\theta)}{m(x_1, x_2)}$$

$$= \frac{f(x_2|\theta)\pi(\theta|x_1)}{\int f(x_2|\theta)\pi(\theta|x_1)\, d\theta}.$$

The last equality follows from

$$\frac{m(x_1)}{m(x_1, x_2)} \int f(x_2|\theta)\pi(\theta|x_1)\, d\theta = \frac{\int f(x_2|\theta)f(x_1|\theta)\pi(\theta)d\theta}{m(x_1, x_2)} = 1.$$

The Bayesian inference is carried out within a probabilistic setting as it is extracted from the posterior distribution $\pi(\theta|\mathbf{x})$. Thus, conditional on the data $\mathbf{x}$, we can compute the posterior probability that θ is in a specific set A as

$$\Pr(A|\mathbf{x}) = \int_{\theta \in A} \pi(\theta|\mathbf{x})\, d\theta.$$

The relationship between the prior $\Pr(A)$ and the posterior $\Pr(A|\mathbf{x})$ is an interesting question. One may wonder what sets have a posterior probability greater than their prior probability. Those sets have been characterized by Piccinato (1984). He proved that for any prior distribution, the posterior probability of a set A is greater than or equal to its prior probability if and only if the set A is a likelihood set $K(g, \mathbf{x})$ for some $g \geq 0$ and $\mathbf{x}$.

Further, confidence sets are now easily interpreted. For instance, for the data $\mathbf{x}$, the Highest Posterior Density (HPD) region of probability $1 - \alpha \in (0, 1)$ is the smallest set in Θ that contains the parameter θ with probability $1 - \alpha$. Thus, it is given by

$$\text{HPD}(k_\alpha) = \{ \theta : \pi(\theta|\mathbf{x}) \geq k_\alpha \},$$

where k_α is the largest value such that

$$\int_{\text{HPD}(k_\alpha)} \pi(\theta|\mathbf{x})\, d\theta = 1 - \alpha.$$

The computation of the HPD(k_α) is quite simple when the parameter space is one–dimensional and the posterior density is unimodal and symmetric with respect to its mode. Otherwise, the computation of the HPD typically requires numerically solving a nonlinear equation.

Example 2.9 (continued). *The 95% HPD for β conditional on the data with $n = 5$ and $\bar{x} = 1$ in Example 2.9 turns out to be* $(7.32, 12.55)$.

Bayesian and MLE estimators might be different for small or moderate sample sizes. Further, Bayesian estimators typically show more stability than MLEs. Let us illustrate this assertion on the estimation of parameter n and θ of the binomial distribution with two unknown parameters in Example 2.8.

Example 2.8 (continued). *Let us assume that θ and n are independent a priori, and that the prior distribution*

$$\pi^J(\theta) = \frac{1}{\pi}\theta^{-1/2}(1 - \theta)^{-1/2}, \quad 0 < \theta < 1,$$

is assigned to θ. This prior is known as the Jeffreys prior and it is the objective prior for doing inference on the parameter θ of a Bernoulli distribution. The notion of objective prior will be introduced in Section 2.3.3. To n we assign the prior

$$\pi(n) = \frac{6}{\pi^2}\frac{1}{n^2}, \quad n = 1, 2, \ldots$$

This prior for n is proper with the mode at point $n = 1$ and a heavy right tail, in fact it does not have a mean. The posterior expectation of n for the sample $\mathbf{x} = (x_1, \ldots, x_N)$ is given by

$$\mathbb{E}(n|\mathbf{x}) = \frac{A}{B},$$

where

$$A = \sum_{n=\max_{i=1,\ldots,N}(x_i)}^{\infty} \frac{1}{n}\frac{\Gamma(Nn - N\bar{x} + 1/2)}{\Gamma(Nn + 1)} \prod_{i=1}^{N}\binom{n}{x_i},$$

$$B = \sum_{n=\max_{i=1,\ldots,N}(x_i)}^{\infty} \frac{1}{n^2}\frac{\Gamma(Nn - N\bar{x} + 1/2)}{\Gamma(Nn + 1)} \prod_{i=1}^{N}\binom{n}{x_i}.$$

Likewise, the posterior expectation of θ given the sample $\mathbf{x}$ is

$$\mathbb{E}(\theta|\mathbf{x}) = \frac{A'}{B'}$$

where

$$A' = \sum_{n=\max_{i=1,\ldots,N}(x_i)}^{\infty} \frac{1}{n^2} \frac{\Gamma(Nn - N\bar{x} + 1/2)\Gamma(N\bar{x} + 3/2)}{\Gamma(Nn + 2)} \prod_{i=1}^{N} \binom{n}{x_i},$$

$$B' = \sum_{n=\max_{i=1,\ldots,N}(x_i)}^{\infty} \frac{1}{n^2} \frac{\Gamma(Nn - N\bar{x} + 1/2)\Gamma(N\bar{x} + 1/2)}{\Gamma(Nn + 1)} \prod_{i=1}^{N} \binom{n}{x_i}.$$

Table 2.2 displays the comparison between the Bayesian and maximum likelihood estimators of the parameters of the Binomial distribution for samples $\mathbf{x}_1 = (16, 18, 22, 25, 27)$ and $\mathbf{x}_2 = (16, 18, 22, 25, 28)$.

TABLE 2.2

Bayesian and maximum likelihood estimators of n and θ of a Binomial distribution for $\mathbf{x}_1$ and $\mathbf{x}_2$

Sample	Bayesian estimator		MLE	
	$\mathbb{E}(n|\mathbf{x}_i)$	$\mathbb{E}(\theta|\mathbf{x}_i)$	$\hat{n}(\mathbf{x}_i)$	$\hat{\theta}(\mathbf{x}_i)$
$\mathbf{x}_1 = (16, 18, 22, 25, 27)$	106	0.375	99	0.218
$\mathbf{x}_2 = (16, 18, 22, 25, 28)$	119	0.358	190	0.114

This example illustrates the fact that the Bayesian estimator of n and θ is much more stable than the MLE.

For large sample sizes, however, Bayesian and MLE estimators are very close to each other, which is a consequence of the following result.

Let $\{x_i, \; i \geq 1\}$ be a sequence of i.i.d. random variables with distribution $f(x|\theta_0)$ for $\theta_0 \in \Theta$. Then, under wide regularity conditions the Bayesian estimator $\mathbb{E}(\theta|x_1, \ldots, x_n)$ and the MLE $\hat{\theta}(x_1, \ldots, x_n)$ are such that for any $\varepsilon > 0$,

$$\Pr\left(|\mathbb{E}(\theta|x_1, \ldots, x_n) - \hat{\theta}(x_1, \ldots, x_n)| \geq \varepsilon\right) = O(n^{-1}),$$

where the probability is computed with the distribution $f(x|\theta_0)$.

Furthermore, from the above equality it follows that the Bayesian estimator $\mathbb{E}(\theta|\mathbf{x})$ is a consistent estimator of θ. Under regularity conditions it can be shown that the sequence of posterior distributions $\{\pi(\theta|x_1, \ldots, x_n), \ n \geq 1\}$ converges in probability, as n tends to infinity, to the degenerate distribution on θ_0. Thus, the sequence of posterior expectations of θ,

$$\mathbb{E}(\theta|x_1, \ldots, x_n) = \frac{\int_\Theta \theta \left(\prod_{i=1}^n f(x_i|\theta)\right) \pi(\theta) \, d\theta}{\int_\Theta \left(\prod_{i=1}^n f(x_i|\theta)\right) \pi(\theta) \, d\theta},$$

tends to θ_0 in probability as n tends to infinity. Regularity conditions for this result to hold are given in Sections 10.9 and 10.10 in DeGroot (1970). Further, it is there shown that a normal distribution with mean at $\hat{\theta}(x_1, \ldots, x_n)$ and variance $1/(nI(\theta))$ approximates the posterior distribution $\pi(\theta|x_1, \ldots, x_n)$ for large n.

For small or moderate sample sizes, the contribution of the prior to the posterior distribution might be relevant, and hence a main question in Bayesian parametric inference is how a prior for the parameter θ of the sampling model $f(x|\theta)$ is elicited. We briefly discuss some approaches to answer this question.

2.3.1 Subjective priors

A distribution $\pi(\theta)$ that incorporates subjective information on the parameter θ is called a *subjective prior*. One may ask what kind of information on θ is accessible to an expert. An instance of such accessible information would be the location of some quantiles of the distribution $\pi(\theta)$. In particular, detection of symmetries of $\pi(\theta)$.

There is an extensive literature on the process of assigning or eliciting subjective probability distributions for the parameters of a model from expert knowledge (Savage, 1971; Murphy and Winkler, 1977; Genest and Zidek, 1986; O'Hagan et al., 2006). Chaloner (1996) provides a review of methods of eliciting prior information in clinical trials, including interviews with clinicians, questionnaires, etc. Other papers on the elicitation process in clinical data are Freedman and Spiegelhalter

(1983); Spiegelhalter and Freedman (1986); Wolpert (1989); Spiegelhalter et al. (1994); Kadane and Wolfson (1995); Spiegelhalter et al. (1995); Chaloner and Rhame (2001); and Vázquez–Polo and Negrín (2004).

It seems to be clear that, in general, information on the tails of $\pi(\theta)$ is hardly an accessible prior information, and hence the tails are the bottleneck of the subjective elicitation. This seems to imply that a class of priors with different tails compatible with the accessible prior information would be a realistic subjective model. Bayesian inference associated with the idea of having a class of priors instead of a single prior was developed in the eighties and nineties, and grouped under the heading "Bayesian Robustness." Interesting thorough reviews of Bayesian Robustness include Berger (1994), Berger et al. (1996) and Rios and Ruggeri (2000).

In Bayesian Robustness the classes of distributions are typically nonparametric classes, and unfortunately they usually provide a non–robust Bayesian inference of our quantity of interest. An exception was given by Wasserman (1989). He proved that the likelihood sets $K(g, \mathbf{x})$ are robust to contamination of the prior $\pi_0(\theta)$. This means that the posterior probability of the likelihood sets is relatively insensitive to the prior when it varies in the class of contaminated priors

$$\Gamma = \{\pi(\theta) : \pi(\theta) = (1 - \varepsilon)\pi_0(\theta) + \varepsilon q(\theta), q \in \mathcal{Q}\}$$

where $\varepsilon \in [0, 1]$ and $\mathcal{Q}$ is the set of all probability distributions.

The lack of robustness always appears when θ is a multidimensional parameter. This yields considering predetermined tails for the prior for θ.

2.3.2 Conjugate priors

A possible solution to the difficulty of eliciting the tails of the prior distribution $\pi(\theta)$ would be to choose the tails as those of the likelihood of θ. This yields the notion of conjugate priors.

Let $\mathbf{x} = (x_1, ..., x_n)$ be a random sample from a sampling distribution in $\mathcal{F} = \{f(x|\theta),\ \theta \in \Theta\}$. A parametric class of priors

$\mathcal{C} = \{\pi(\theta|\lambda), \ \lambda \in \Lambda\}$ is called a *conjugate class of priors* with respect to the sampling model $f(x|\theta)$ in $\mathcal{F}$ if for any prior $\pi(\theta|\lambda)$ in $\mathcal{C}$ the posterior distribution $\pi(\theta|\mathbf{x})$ is also in $\mathcal{C}$ for any $\mathbf{x}$.

Example 2.1 (continued). *For the Bernoulli sampling family given in (2.11), the conjugate class of priors is the beta family*

$$\mathcal{C} = \{\pi(\theta|\alpha, \beta), \ \alpha, \beta > 0\},$$

that is

$$\pi(\theta|\alpha, \beta) = \frac{\Gamma(\alpha + \beta)}{\Gamma(\alpha)\Gamma(\beta)} \, \theta^{\alpha-1}(1-\theta)^{\beta-1}, \quad \theta \in (0, 1). \qquad (2.37)$$

For, the posterior distribution of θ is

$$
\begin{aligned}
\pi(\theta|\mathbf{x}) \ &= \ \frac{\ell_{\mathbf{x}}(\theta)\pi(\theta)}{\int_0^1 \ell_{\mathbf{x}}(\theta)\pi(\theta)\, d\theta} \\
&= \ \frac{\Gamma(\alpha + \beta + n)}{\Gamma(n\bar{x} + \alpha)\Gamma(n - n\bar{x} + \beta)} \, \theta^{n\bar{x}+\alpha-1}(1-\theta)^{n-n\bar{x}+\beta-1},
\end{aligned}
$$

$$(2.38)$$

which is a beta distribution with updated parameters $n\bar{x} + \alpha$ and $n - n\bar{x} + \beta$.

Example 2.2 (continued). *For the Poisson sampling model given in (2.13) the conjugate class of priors is the gamma family*

$$\mathcal{C} = \{\pi(\lambda|\alpha, \beta), \ \alpha, \beta > 0\},$$

that is

$$\pi(\lambda|\alpha, \beta) = \frac{\beta^{\alpha}}{\Gamma(\alpha)} \, \lambda^{\alpha-1} \exp(-\beta\lambda), \quad \lambda > 0.$$

Indeed, the posterior distribution of λ is

$$
\begin{aligned}
\pi(\lambda|\mathbf{x}) \ &= \ \frac{\ell_{\mathbf{x}}(\lambda)\pi(\lambda)}{\int_0^\infty \ell_{\mathbf{x}}(\lambda)\pi(\lambda)\, d\lambda} \\
&= \ \frac{(n + \beta)^{n\bar{x}+\alpha}}{\Gamma(n\bar{x} + \alpha)} \lambda^{n\bar{x}+\alpha-1} \exp(-(n + \beta)\lambda),
\end{aligned}
$$

which is a gamma distribution with updated parameters $n\bar{x} + \alpha$ and $n + \beta$.

Example 2.10 (Normal sampling model with known variance). *For the normal sampling model given in (2.15) with known variance $\mathcal{N}(x|\mu, \sigma_0^2)$, the conjugate class of priors for the parameter μ is the normal family*

$$\mathcal{C} = \{\mathcal{N}(\mu|\mu_1, \sigma_1^2), \ \mu_1 \in \mathbb{R}, \sigma_1 \in \mathbb{R}^+\}.$$

Indeed, for the sample $\mathbf{x}$ the posterior distribution of μ is the normal family

$$\pi(\mu|\mathbf{x}) = \mathcal{N}(\mu|\mu_2, \sigma_2^2),$$

where

$$\mu_2 = \mu_1 \frac{\sigma_0^2}{\sigma_0^2 + n\sigma_1^2} + \bar{x} \frac{n\sigma_1^2}{\sigma_0^2 + n\sigma_1^2}, \quad \sigma_2^2 = \left(\frac{1}{\sigma_1^2} + \frac{n}{\sigma_0^2}\right)^{-1}, \quad (2.39)$$

with $\bar{x} = \sum_{i=1}^n x_i/n$. We note that the posterior mean is a convex combination of the prior mean μ_1 and the sample mean $\bar{x}$.

We remark that hyperparameters μ_2 and σ_2 in the normal example can be accommodated to match accessible subjective information on μ, and the same can be said for the hyperparameters α and β of the preceding examples.

Most of the one–dimensional models considered so far can be written as a member of the exponential family

$$f(x|\theta) = g(x)h(\theta) \exp\left(t(x)\psi(\theta)\right). \quad (2.40)$$

For a random sample $\mathbf{x} = (x_1, \ldots, x_n)$ the likelihood function of θ is

$$\ell_{\mathbf{x}}(\theta) \propto h(\theta)^n \exp\left\{\sum_{i=1}^n t(x_i)\psi(\theta)\right\}, \quad (2.41)$$

and thus, $\sum_{i=1}^n t(x_i)$ is the sufficient statistics for θ, a result that follows from the Halmos–Savage factorization theorem (Zacks, 1981). The conjugate class of priors for θ is

$$\mathcal{C} = \{\pi(\theta|\alpha, \beta), \ \alpha, \beta \in \mathbb{R}\},$$

where

$$\pi(\theta|\alpha, \beta) \propto h(\theta)^\alpha \exp\left(\beta\psi(\theta)\right). \tag{2.42}$$

In fact, from (2.41) and (2.42), the posterior distribution for θ is given by

$$\pi(\theta|\mathbf{x}) \propto h(\theta)^{n+\alpha} \exp\left\{ \left(\sum_{i=1}^{n} t(x_i) + \beta \right) \psi(\theta) \right\},$$

which is also in $\mathcal{C}$.

The one–parameter exponential family can be extended to the k–parametric exponential family

$$f(x|\boldsymbol{\theta}) = g(x)h(\boldsymbol{\theta}) \exp\left(\sum_{j=1}^{k} t_j(x)\psi_j(\boldsymbol{\theta}) \right).$$

Then, the conjugate priors are given by

$$\pi(\boldsymbol{\theta}) \propto h(\boldsymbol{\theta})^\alpha \exp\left\{ \sum_{j=1}^{k} \beta_j \psi_j(\boldsymbol{\theta}) \right\}.$$

A detailed analysis of conjugate priors is given in DeGroot (1970) and an extensive catalog of conjugate classes of priors in Bernardo and Smith (1994), pp. 436–442.

2.3.3   Objective priors

When the prior information on θ is weak, an option that has many followers in the statistical community is that of using the so–called *objective priors*. These are operational priors derived with the help of the sampling model $f(x|\theta)$. A first objective method for constructing objective priors is due to Jeffreys (1961). For a real parameter θ in the model $f(x|\theta)$, the Jeffreys prior is given by the squared root of the Fisher information of the model, that is,

$$\pi^J(\theta) \propto \sqrt{I(\theta)}.$$

When θ is a multidimensional parameter, $I(\theta)$ is replaced by the determinant of the Fisher information matrix, although the priors so obtained present some difficulties.

Another method to construct objective priors, called *reference priors*, is due to Bernardo (1979) and further developed in Bernardo and Smith (1994) and Berger et al. (2009), among others. When the parameter θ is one–dimensional, the reference priors $\pi(\theta)$ coincide with the Jeffreys prior $\pi^J(\theta)$.

For a recent extensive review on objective priors, see the paper by Consonni et al. (2018).

A difficulty with objective priors is that for most of the parametric sampling models they are improper, that is, they do not integrate a finite quantity. This implies that the reference prior cannot typically be normalized to a probability distribution, and hence it is well defined up to an arbitrary positive multiplicative constant.

While this is not a difficulty for computing the posterior distribution of θ, as the arbitrary constant cancels out in the ratio appearing in the posterior distribution, it is a serious drawback when the inference is based on the marginal of the sample $m(\mathbf{x}) = \int_\Theta f(\mathbf{x}|\theta)\pi(\theta)\,d\theta$ as occurs in hypotheses testing. Therefore, improper priors are unsuitable for model selection or hypotheses testing. Let us illustrate this fact on a normal example with known variance.

Example 2.10 (continued). *For the family of normal sampling models with known variance, for simplicity $\sigma_0^2 = 1$, $\{\mathcal{N}(x|\theta, 1), \theta \in \mathbb{R}\}$, the Jeffreys prior for θ is given by*

$$\pi^J(\theta) = c\,\mathbf{1}_{(-\infty,\infty)}(\theta),$$

where c is an arbitrary positive constant. The posterior distribution of θ for the random sample $\mathbf{x} = (x_1, \ldots, x_n)$ is the well–defined normal probability density with mean $\bar{x} = \sum_{i=1}^n x_i/n$ and variance $1/n$, that is,

$$\pi(\theta|\mathbf{x}) = \mathcal{N}\left(\theta|\bar{x}, \frac{1}{n}\right).$$

However, the marginal distribution of the sample $\mathbf{x}$ is given by

$$m(\mathbf{x}) = c \, \frac{1}{n^{1/2}(2\pi)^{(n-1)/2}} \, \exp\left\{ -\frac{1}{2} \sum_{i=1}^{n} (x_i - \bar{x})^2 \right\},$$

which is defined up to the arbitrary multiplicative constant c.

2.4   The predictive distribution

In cost–effectiveness analysis the predictive distribution of the cost and the effectiveness play a central role. In general, given a random variable X with distribution in the class $\{f(x|\theta), \theta \in \Theta\}$, and a sample $\mathbf{x} = (x_1, \ldots, x_n)$ from a model in this class, the distribution of a future observation y conditional on the sample is called the *predictive distribution* of the random variable X.

From the frequentist viewpoint the predictive distribution of X is given by $f(y|\hat{\theta})$, where $\hat{\theta} = \hat{\theta}(\mathbf{x})$ is the MLE of θ. The expected value of y is given by

$$\mathbb{E}(Y|\hat{\theta}) = \int y f(y|\hat{\theta}) \, dy.$$

From the Bayesian viewpoint the predictive distribution of X is given by

$$f(y|\mathbf{x}) = \int_{\Theta} f(y|\theta)\pi(\theta|\mathbf{x}) \, d\theta, \tag{2.43}$$

which depends not only on the sampling model $f(x|\theta)$, but also on the prior distribution $\pi(\theta)$ of parameter θ. The expected value of the predictive distribution is given by

$$
\begin{aligned}
\mathbb{E}(Y|\mathbf{x}) &= \int y \, f(y|\mathbf{x}) \, dy \\
&= \int y \left(\int f(y|\theta)\pi(\theta|\mathbf{x}) \, d\theta \right) dy \\
&= \int \left(\int y f(y|\theta) \, dy \right) \pi(\theta|\mathbf{x}) \, d\theta \\
&= \int \mathbb{E}(Y|\theta)\pi(\theta|\mathbf{x}) \, d\theta.
\end{aligned}
\tag{2.44}
$$

Example 2.9 (continued). *Consider the gamma sampling model with a known shape parameter. From (2.43), the Bayesian predictive distribution is*

$$
\begin{aligned}
f(y|\mathbf{x}) &= \int_0^\infty f(y|\beta)\,\pi(\beta|\mathbf{x})\,d\beta \\
&= \frac{\Gamma(\alpha_0(n+1)+\alpha_1)\,\beta_2^{\alpha_2}}{\Gamma(\alpha_2)\Gamma(\alpha_0)}\,\frac{y^{\alpha_0-1}}{(y+\beta_2)^{\alpha_0(n+1)+\alpha_1}}, \quad (2.45)
\end{aligned}
$$

where $\alpha_2 = \alpha_0 n + \alpha_1$ and $\beta_2 = n\bar{x} + \beta_1$.

From (2.44), the expected value of a new observation is

$$
\begin{aligned}
\mathbb{E}(Y|\mathbf{x}) &= \int_0^\infty \frac{\alpha_0}{\beta}\,\pi(\beta|\mathbf{x})\,d\beta \\
&= \alpha_0 \frac{\beta_2^{\alpha_2}}{\Gamma(\alpha_2)}\,\frac{\Gamma(\alpha_2-1)}{\beta_2^{\alpha_2-1}} = \frac{\alpha_0\beta_2}{\alpha_2-1}.
\end{aligned}
$$

Under regularity conditions and for large sample size n, the Bayesian and the frequentist predictive distributions $f(y|\mathbf{x})$ and $f(y|\hat{\theta}(\mathbf{x}))$ are close each other. However, for small sample sizes these distributions differ as the following simple example shows.

Example 2.11. *Suppose X is a binary random variable with Bernoulli distribution $Be(x|\theta) = \theta^x(1-\theta)^{1-x}$ for $x = 0,1$ and $0 \le \theta \le 1$. The random sample $\mathbf{x} = (x_1,\dots,x_n)$ from $Be(x|\theta)$ has distribution*

$$
f(\mathbf{x}|\theta) = \theta^{n\bar{x}}(1-\theta)^{n-n\bar{x}},
$$

where $n\bar{x} = \sum_{i=1}^n x_i$. The maximum likelihood estimator of θ is $\hat{\theta}(\bar{x}) = \bar{x}$, and the frequentist predictive distribution

$$
\widehat{\Pr}(y|\bar{x}) = \bar{x}^y(1-\bar{x})^{1-y}, \; y = 0,1.
$$

On the other hand, the Jeffreys objective prior for θ is the proper Beta distribution with parameters $(1/2,1/2)$, that is

$$
\pi^J(\theta) = \frac{1}{\pi}\theta^{-1/2}(1-\theta)^{-1/2}, \; 0 < \theta < 1.
$$

The posterior distribution of θ conditional on the sample is given by

$$
\pi(\theta|\bar{x},n) = \frac{\Gamma(n+1)}{\Gamma(1/2+n-n\bar{x})\Gamma(1/2+n\bar{x})}\theta^{n\bar{x}-1/2}(1-\theta)^{n-n\bar{x}-1/2}.
$$

Then, the Bayesian predictive distribution for the Jeffreys prior is given by the probability function

$$\Pr(y|\bar{x}, n) = \frac{\Gamma(3/2 + n - n\bar{x} - y)\Gamma(1/2 + n\bar{x} + y)}{(n + 1)\Gamma(1/2 + n - n\bar{x})\Gamma(1/2 + n\bar{x})}, \quad y = 0, 1.$$

Values of the Bayesian and frequentist predictive distributions for $\bar{x} = 0.2$ and $n = 5, 10, 15, 20, 25, 30$ are given in the second and third rows, in Table 2.3.

TABLE 2.3

Bayesian predictive probability for $y = 1$ (second row), and frequentist predictive probability for $y = 1$ (third row).

$(\bar{x}, n)$	$(0.2, 5)$	$(0.2, 10)$	$(0.2, 15)$	$(0.2, 20)$	$(0.2, 25)$	$(0.2, 30)$
$\Pr(y = 1|\bar{x}, n)$	0.25	0.23	0.22	0.21	0.21	0.21
$\widehat{\Pr}(y = 1|\bar{x})$	0.20	0.20	0.20	0.20	0.20	0.20

These numbers show that the difference between these predictive distributions is large when the sample size n is small, and the difference decreases as n increases.

We observe that the frequentist predictive distribution depends on the MLE but it does not explicitly depend on the sample size n. However, $\widehat{\Pr}(y|\bar{x})$ and $\Pr(y|\bar{x}, n)$ have a consistent asymptotic behavior. When sampling from $Be(x|\theta_0)$ we have that

$$\lim_{n \to \infty} \widehat{\Pr}(y = 1|\bar{x}) = \lim_{n \to \infty} \bar{x} = \theta_0, \qquad [P_{\theta_0}],$$

and

$$\lim_{n \to \infty} \Pr(y = 1|\bar{x}, n) = \lim_{n \to \infty} \frac{\frac{3}{2} n^{-1} + \bar{x}}{1 + 3n^{-1}} = \lim_{n \to \infty} \bar{x} = \theta_0, \quad [P_{\theta_0}].$$

The Bayesian predictive distribution depends on the prior distribution we are using, and an alternative to the Jeffreys prior is the uniform prior $\pi^U(\theta) = \mathbf{1}_{(0,1)}(\theta)$, which was originally utilized by Bayes

(1763). However, the predictive distribution for the uniform prior and the Jeffreys prior are very close each other. Other priors can be used, although the Jeffreys and the uniform are by far the most utilized priors for binary data.

There are more sophisticated versions of the frequentist predictive distribution that we do not bring here. References on this topic include Komaki (1996) and Corcuera and Giummolè (1999).

2.5   Bayesian model selection

Let $M_1 = \{f(x|\theta_1), \theta_1 \in \Theta_1 \subset \mathbb{R}^k\}$ be a class of sampling models for the random variable X, and $\mathbf{x} = (x_1, \ldots, x_n)$ a sample from an unknown distribution in the class. The integer k is called the *dimension of the model*, and an important question is whether this dimension can be reduced on the basis of the sampling information $\mathbf{x}$, that is, whether the set of models M_1 can be restricted to a subset $M_0 = \{f(x|\theta_0), \theta_0 \in \Theta_0\}$, where Θ_0 is a specified subset of Θ_1. The class of sampling models M_0 is said to be *nested* in M_1, that is, the equality $f(x|\theta_1) = f(x|\theta_0)$ if $\theta_1 = \theta_0$ holds.

This decision problem is called *hypothesis testing* or *model selection*. The reduced class of sampling model M_0 is denoted as the *null hypothesis* and the full class M_1 as the *alternative hypothesis*.

The Bayesian formulation of these models needs prior distributions for the model parameters and models, and hence the Bayesian model can be written as

$$M_0 : \{f(\mathbf{x}|\theta_0, M_0), \pi(\theta_0, M_0)\},$$

and

$$M_1 : \{f(\mathbf{x}|\theta_1, M_1), \pi(\theta_1, M_1)\},$$

where $f(\mathbf{x}|\theta_j, M_j) = \prod_{i=1}^{n} f(x_i|\theta_j, M_j)$, $j = 0, 1$.

In model selection we are not interested in estimating θ_0 or θ_1, but we are interested in estimating model M_0 and M_1 based on the information given by a sample $\mathbf{x}$ drawn from a sampling model in M_1. The usual model estimator is the model having the highest posterior probability in the model space $\{M_0, M_1\}$.

For computing the posterior model probabilities we decompose the priors as

$$\pi(\theta_0, M_0) = \pi(\theta_0|M_0)\pi(M_0)$$

and

$$\pi(\theta_1, M_1) = \pi(\theta_1|M_1)\pi(M_1),$$

where $\pi(M_0) + \pi(M_1) = 1$. Then, the distribution of the sample $\mathbf{x}$ conditional on M_0 is given by

$$m(\mathbf{x}|M_0) = \int_{\Theta_0} f(\mathbf{x}|\theta_0, M_0)\pi(\theta_0|M_0)\, d\theta_0,$$

and on M_1 by

$$m(\mathbf{x}|M_1) = \int_{\Theta_1} f(\mathbf{x}|\theta_1, M_1)\pi(\theta_1|M_1)\, d\theta_1.$$

The marginal distributions $m(\mathbf{x}|M_0)$ and $m(\mathbf{x}|M_1)$ are the likelihoods of M_0 and M_1 for the sample $\mathbf{x}$.

From the Bayes theorem, the posterior probabilities of M_0 and M_1 in the model space $\{M_0, M_1\}$ turn out to be

$$\mathrm{Pr}(M_0|\mathbf{x}) = \frac{m(\mathbf{x}|M_0)\pi(M_0)}{m(\mathbf{x}|M_0)\pi(M_0) + m(\mathbf{x}|M_1)\pi(M_1)},$$

and

$$\mathrm{Pr}(M_1|\mathbf{x}) = 1 - \mathrm{Pr}(M_0|\mathbf{x}).$$

The posterior model probability of M_0 is usually written as

$$\mathrm{Pr}(M_0|\mathbf{x}) = \frac{\pi(M_0)}{\pi(M_0) + B_{10}(\mathbf{x})\pi(M_1)},$$

where

$$B_{10}(\mathbf{x}) = \frac{m(\mathbf{x}|M_1)}{m(\mathbf{x}|M_0)}$$

is the model likelihood ratio, and it is called the *Bayes factor* to compare model M_1 *versus* M_0. We note that the Bayes factor contains all the sampling information for model selection, and it does not depend on the prior probabilities of the models.

The objective prior distribution typically used for the models $\{M_0, M_1\}$ is the uniform prior

$$\pi(M_0) = \pi(M_1) = \frac{1}{2}.$$

To simplify, $\pi(\theta_j | M_j)$ will be denoted by $\pi(\theta_j)$ for $j = 0, 1$.

Unfortunately, the objective reference priors for the model parameters $\pi(\theta_0)$ and $\pi(\theta_1)$ are typically improper, and hence the Bayes factor for those priors is defined up to an arbitrary positive multiplicative constant. Thus, the posterior model probabilities are defined up to an arbitrary positive constant.

A solution to this problem consists of replacing the objective reference priors with the so-called *intrinsic priors for model selection* that we introduce in the next section.

2.5.1 Intrinsic priors for model selection

A way of converting improper priors for model parameters $\pi(\theta_0)$ and $\pi(\theta_1)$ into priors suitable for model selection was given by Berger and Pericchi (1996a) and further explored by Moreno (1997) and Moreno et al. (1998). For a review of model selection see Lahiri (2001).

The so-called *intrinsic priors for model selection* are constructed in two steps. In a first step the intrinsic prior for θ_1, conditional on θ_0, is defined as

$$\pi^I(\theta_1 | \theta_0, m) = \pi(\theta_1) \mathbb{E}_{\mathbf{y}|\theta_1, M_1} \frac{f(\mathbf{y}|\theta_0, M_0)}{m(\mathbf{y}|M_1)}, \qquad (2.46)$$

where the random vector $\mathbf{y} = (y_1, \dots, y_m)$ is the *training sample*, m is the *training sample size*, $f(\mathbf{y}|\theta_0, M_0) = \prod_{j=1}^{m} f(y_j|\theta_0, M_0)$, the expectation is taken with respect to $f(\mathbf{y}|\theta_1, M_1) = \prod_{j=1}^{m} f(y_j|\theta_1, M_1)$, and

$$m(\mathbf{y}|M_1) = \int_{\Theta_1} f(\mathbf{y}|\theta_1, M_1)\pi(\theta_1) \, d\theta_1.$$

It is easy to show that $\pi^I(\theta_1|\theta_0, m)$ is a probability density for any training sample size m such that

$$0 < \int_{\Theta_1} f(\mathbf{y}|\theta_1, M_1)\pi(\theta_1)\, d\theta_1 < \infty.$$

In fact,

$$\begin{aligned}
\int_{\Theta_1} \pi^I(\theta_1|\theta_0, m)\, d\theta_1 &= \int_{\Theta_1} \pi(\theta_1) \left(\int_{\mathcal{X}^m} \frac{f(\mathbf{y}|\theta_0, M_0)f(\mathbf{y}|\theta_1, M_1)}{m(\mathbf{y}|M_1)} d\mathbf{y} \right) d\theta_1 \\
&= \int_{\mathcal{X}^m} \frac{f(\mathbf{y}|\theta_0, M_0)}{m(\mathbf{y}|M_1)} \left(\int_{\Theta_1} f(\mathbf{y}|\theta_1, M_1)\pi(\theta_1)\, d\theta_1 \right) d\mathbf{y} \\
&= \int_{\mathcal{X}^m} \frac{f(\mathbf{y}|\theta_0, M_0)}{m(\mathbf{y}|M_1)} m(\mathbf{y}|M_1)\, d\mathbf{y} \\
&= \int_{\mathcal{X}^m} f(\mathbf{y}|\theta_0, M_0)\, d\mathbf{y} = 1.
\end{aligned}$$

In a second step the unconditional intrinsic prior for θ_1 is obtained from $\pi^I(\theta_1|\theta_0, m)$ integrating out θ_0, that is,

$$\pi^I(\theta_1|m) = \int_{\Theta_0} \pi^I(\theta_1|\theta_0, m)\pi(\theta_0)\, d\theta_0.$$

The pair of priors $\{\pi(\theta_0), \pi^I(\theta_1|m)\}$ is called the intrinsic priors for model selection in the model space $\{M_0, M_1\}$. It is easy to see that even when the unconditional intrinsic priors are defined up to the arbitrary multiplicative constant c_0, the posterior model probability $\Pr(M_0|\mathbf{x})$ and $\Pr(M_1|\mathbf{x})$ for intrinsic priors are well defined.

Model selection for intrinsic priors has been proved to enjoy excellent behavior for moderate and large sample sizes (Casella and Moreno, 2006; Moreno et al., 2015, and references therein). Therefore, for model selection, intrinsic priors are strongly recommended.

The intrinsic priors depend on the training sample size m and the problem of choosing it has been considered by Berger and Pericchi (1996a), Casella and Moreno (2006), and León–Novelo et al. (2012), among others. The original proposal by Berger and Pericchi was to choose m with a minimal size such that the marginals $m(\mathbf{y}|M_0)$ and $m(\mathbf{y}|M_1)$ are finite and greater than zero.

Another proposal consists of choosing m such that the conditional intrinsic prior $\pi^I(\theta_1|\theta_0, m)$ satisfies a given condition to be fixed for the specific problem. For instance, in a meta–analysis context, a specific value of the linear correlation between the parameters θ_1 and θ_0 was imposed (Moreno et al., 2014).

Let us illustrate, on a normal example, the construction of the intrinsic priors.

Example 2.12. *Let $M_1 : \{\mathcal{N}(x|\mu, \tau^2), (\mu, \tau) \in \mathbb{R} \times \mathbb{R}^+\}$ be the family of normal sampling models with unknown mean and variance. Suppose that we are interested in the model selection between model M_1 and model $M_0 : \{\mathcal{N}(x|\mu_0, \sigma^2), \sigma \in \mathbb{R}^+\}$ where μ_0 is a fixed point. This model selection problem is equivalent to testing the null $H_0 : \mu = \mu_0$ versus $H_1 : \mu \in \mathbb{R}$.*

The Bayesian formulation of this problem is the model selection between model

$$M_0 : \{\mathcal{N}(x|\mu_0, \sigma^2), \pi(\sigma)\},$$

and

$$M_1 : \{\mathcal{N}(x|\mu, \tau^2), \pi(\mu, \tau)\},$$

where

$$\pi(\sigma) = \frac{c_0}{\sigma} \mathbf{1}_{\mathbb{R}^+}(\sigma),$$

and

$$\pi(\mu, \tau) = \frac{c_1}{\tau} \mathbf{1}_{\mathbb{R} \times \mathbb{R}^+}(\mu, \tau),$$

are the reference priors for σ and (μ, τ), and c_0 and c_1 are arbitrary positive constants. We also assume that

$$\pi(M_0) = \pi(M_1) = \frac{1}{2}.$$

Applying the expression (2.46) for the training sample size $m = 2$, the intrinsic prior for (μ, τ), conditional on (μ_0, σ), turns out to be

$$\pi^I(\mu, \tau|\mu_0, \sigma, m = 2) = \mathcal{N}\left(\mu|\mu_0, \frac{\tau^2 + \sigma^2}{2}\right) HC^+(\tau|0, \sigma),$$

where $HC^+(\tau|0,\sigma)$ represents the half Cauchy distribution with location parameter 0 and scale parameter σ, that is,

$$HC^+(\tau|0,\sigma) = \frac{2}{\pi}\frac{\sigma}{\tau^2 + \sigma^2}.$$

Integrating out σ in $\pi^I(\mu,\tau|\mu_0,\sigma)$ with respect to $\pi(\sigma)$, the unconditional intrinsic prior turns out to be

$$\begin{aligned}
\pi^I(\mu,\tau|\mu_0,m=2) &= \int_0^\infty \pi^I(\mu,\tau|\mu_0,\sigma,m=2)\pi(\sigma)\,d\sigma \\
&= c_0\frac{2}{\pi}\int_0^\infty \mathcal{N}\left(\mu|\mu_0,\frac{\tau^2+\sigma^2}{2}\right)\frac{1}{\tau^2+\sigma^2}\,d\sigma.
\end{aligned}$$

If we substitute the reference prior $\pi(\mu,\tau)$ with the intrinsic prior $\pi^I(\mu,\tau|\mu_0,m=2)$, the Bayesian model selection problem becomes that of choosing between model

$$M_0 : \left\{\mathcal{N}(x|\mu_0,\sigma^2), \pi(\sigma)\right\}$$

and model

$$M_1 : \left\{\mathcal{N}(x|\mu,\tau^2), \pi^I(\mu,\tau|\mu_0,m=2)\right\}.$$

This problem has a well–defined solution. Indeed, for the sample $\mathbf{x} = (x_1,\ldots,x_n)$ from a unknown model in M_1, the marginal of the data $\mathbf{x}$, conditional on M_0, turns out to be

$$m(\mathbf{x}|M_0) = \frac{1}{2\pi^{n/2}}\frac{\Gamma(n/2)}{(n(s^2+(\bar{x}-\mu_0)^2))^{n/2}} \tag{2.47}$$

and conditional on M_1

$$m(\mathbf{x}|M_1) = \frac{2^{n/2}\Gamma(n/2)}{(\sqrt{2\pi})^n\sqrt{n}\,\pi}I(\bar{x},s^2,n),$$

where

$$I(\bar{x},s^2,n) = \int_0^{\pi/2} K_1(\varphi,\bar{x},s^2,n)K_2(\varphi,\bar{x},s^2,n)\,d\varphi,$$

with

$$K_1(\varphi,\bar{x},s^2,n) = \frac{1}{\sin^{n-1}\varphi\left(\frac{n+2}{2n}\sin^2\varphi + \frac{1}{2}\cos^2\varphi\right)^{1/2}},$$

and

$$K_2(\varphi, \bar{x}, s^2, n) = \left(\frac{(\bar{x} - \mu_0)^2}{\frac{n+2}{2n} \sin^2 \varphi + \frac{1}{2} \cos^2 \varphi} + \frac{ns^2}{\sin^2 \varphi} \right)^{-n/2}.$$

The integral on the interval $(0, \pi/2)$ has no close form but numerically it is easily solved.

Thus, the posterior probability of model M_0 is given by

$$\Pr(M_0|\mathbf{x}) = \frac{1}{1 + B_{10}(\mathbf{x})}, \tag{2.48}$$

where

$$
\begin{aligned}
B_{10}(\mathbf{x}) &= \frac{m(\mathbf{x}|M_1)}{m(\mathbf{x}|M_0)} \\
&= \frac{2I(\bar{x}, s^2, n)}{\pi \sqrt{n}} [ns^2 + n(\bar{x} - \mu_0)^2]^{n/2}. \tag{2.49}
\end{aligned}
$$

We use simulated data from a normal distribution to illustrate the performance of the Bayesian model selection for the intrinsic priors in Example 2.12. We also compared the conclusions from this Bayesian procedure with those from the $p-$values of the Student t test.

Example 2.12 (continued). *We fix $\mu_0 = 0$ and hence the Bayesian models to be compared are*

$$M_0 : \left\{ \mathcal{N}(x|0, \sigma^2), \ \pi(\sigma) = \frac{c_0}{\sigma} \right\},$$

and

$$M_1 : \left\{ \mathcal{N}(x|\mu, \tau^2), \ \mathcal{N}\left(\mu|0, \frac{\tau^2 + \sigma^2}{2}\right) HC^+(\tau|0, \sigma) \right\}.$$

We assume the model prior

$$\pi(M_0) = \pi(M_1) = \frac{1}{2},$$

and compute the posterior probability of model M_0 given in (2.48) for a simulated sample of size $n = 30$ from the null distribution $\mathcal{N}(x|0, 1)$. For these data we also compute the $p-$value for the testing problem

$$H_0 : \mu = 0 \ versus \ H_1 : \mu \in \mathbb{R}$$

using the well–known t test.

We repeat the simulation 100 times and compute the mean across simulations of the posterior probability of model M_0 and the mean of the p–values. These mean values are displayed in the third row of Table 2.4.

On the other hand, we do the same calculations for simulated samples from the alternative model $\mathcal{N}(x|1,1)$, and the mean across simulations of the posterior probability of model M_0 and the mean of the p–values are displayed in the last row of Table 2.4.

From Table 2.4 it follows that when sampling from a null model, both the Bayesian and the frequentist procedures accept the null model. The uncertainty of this acceptance is given by a posterior probability of the null as large as 0.8, and a large p–value equal to 0.44. On the other hand, when sampling from an alternative model, the conclusion from the Bayesian and frequentist procedures differ. While the Bayesian procedure rejects the null with probability close to 1, the frequentist procedure accepts the null with a p–value as large as 0.16.

TABLE 2.4

Mean of the posterior probability of model M_0 and p–values across simulations.

Sampling from the null $\mathcal{N}(x|0,1)$	
$\mathrm{Pr}(M_0|\mathbf{x})$	p–value
0.80	0.44

Sampling from the alternative $\mathcal{N}(x|1,1)$	
$\mathrm{Pr}(M_0|\mathbf{x})$	p–value
≈ 0.00	0.16

The discrepancy between the conclusions from the Bayesian and the frequentist procedures for sharp null hypothesis has been well documented in the literature (Berger, 1985). Berger assumed that $\tau = \sigma$

and compared p−values from the normal distribution for specific values of

$$z = \sqrt{n}\,\frac{|\bar{x}|}{\tau},$$

with the posterior probability of M_0 as a function of τ and n. He observed that while samples for which the null is rejected with a given p−value, for instance for $z = 1.96$ the p−value is 0.05, the posterior probability of the null for any prior distribution on μ and any τ do not support this conclusion.

For the case of testing a point null hypothesis from a normal population with known variance, Berger (1985), page 151, Table 4.2, observes that frequentist tests tend to reject the null hypothesis even when its posterior probability is quite substantial (for large samples, $n \geq 50$, this probability is greater than 0.5).

Table 2.4 illustrates that the discrepancy also persists for not necessarily sharp null hypotheses.

The main justification of the intrinsic priors has been to convert improper priors into priors suitable for model selection. However, the intrinsic priors are also useful when the reference priors are not improper. Intrinsic priors for proper priors have been used for testing the equality of two correlated proportions (Consonni and La Rocca, 2008), for studying Bayesian robustness in contingency testing problems (Casella and Moreno, 2009), in the Hardy–Weinberg equilibrium testing problem (Consonni et al., 2011), for comparing nested models for discrete data (Consonni et al., 2013) and for constructing copulas in Bayesian meta–analysis (Moreno et al., 2014).

Extension

Many applications of model selection involve classes of sampling models $\mathcal{M} = \{f(x|\theta_0, M_0), \ldots, f(x|\theta_k, M_k)\}$ with a number of models k greater than one. Variable selection in linear regression and clustering are two important examples of model selection problems in which the number of models k is larger than 1. The extension of the Bayesian formulation of the model selection problem to this case is immediate. The Bayesian models are

$$M_i : \{f(x|\theta_i, M_i), \pi(\theta_i|M_i)\pi(M_i)\}, \quad i = 0, \ldots, k,$$

where the prior for model satisfies $\sum_{i=0}^{k} \pi(M_i) = 1$. For a sample $\mathbf{x} = (x_1, \ldots, x_n)$ from an unknown model in $\mathcal{M}$, the likelihood of model M_i is given by

$$m(\mathbf{x}|M_i) = \int f(\mathbf{x}|\theta_i, M_i)\pi(\theta_i|M_i) \, d\theta_i,$$

and the posterior probability of M_i in $\mathcal{M}$ by

$$\Pr(M_i|\mathbf{x}) = \frac{m(\mathbf{x}|M_i)\pi(M_i)}{\sum_{j=0}^{k} m(\mathbf{x}|M_j)\pi(M_j)}, \quad i = 0, \ldots, k.$$

The model selection rule is again to choose the model with the highest posterior probability.

This Bayesian procedure enjoys very good statistical properties when there is one model in the class $\mathcal{M}$ that is nested in the rest of the models. If we assume that model M_0 is nested in M_i for any $i = 1, \ldots, k$, it is convenient to write the posterior model probabilities as

$$\Pr(M_i|\mathbf{x}) = \frac{B_{i0}(\mathbf{x}) \, \pi(M_i)/\pi(M_0)}{1 + \sum_{j=1}^{k} B_{j0}(\mathbf{x})\pi(M_j)/\pi(M_0)}, \quad i = 0, \ldots, k,$$

where

$$B_{i0}(\mathbf{x}) = \frac{m(\mathbf{x}|M_i)}{m(\mathbf{x}|M_0)}$$

is the Bayes factor for comparing model M_i against M_0. We note that we have written the posterior model probabilities in terms of Bayes factors for nested models. As a consequence, we can prove, for instance, the consistency of the Bayes factor $B_{i0}(\mathbf{x})$ for regular prior distributions for the model parameters when sampling from either M_i or M_0. Further, when sampling from $f(x|\theta_i, M_i)$ it follows that the posterior model probabilities are consistent, that is,

$$\lim_{n\to\infty} \Pr(M_s|\mathbf{x}) = \begin{cases} 1, & \text{if } s = i, \\ 0, & \text{if } s \neq i \end{cases} \qquad [P_{\theta_i}].$$

When the prior for the model parameters $\pi(\theta_i|M_i)$ is improper and thus it cannot be used for computing the likelihood of the model $m(\mathbf{x}|M_i)$, $i = 0, \ldots, k$, as typically happens for the reference priors, the improper priors $(\pi(\theta_0|M_0), \pi(\theta_i|M_i))$ are replaced by the intrinsic priors $(\pi(\theta_0|M_0), \pi^I(\theta_i|\theta_0, m))$. The conditional intrinsic prior $\pi^I(\theta_i|\theta_0, m)$ for $i = 1, \ldots, k$ is computed using expression (2.47). The resulting posterior model probabilities enjoy very good statistical properties (Moreno et al., 2015). Examples of this intrinsic priors construction will be given for clustering in Chapter 5 and for variable selection in Chapter 6.

2.6 The normal linear model

In this section we introduce the normal linear model, and a summary of the frequentist and Bayesian estimations of their parameters. This is a central model in subgroup cost–effectiveness analysis that we develop in Chapter 6.

Definition 2.3. *The random variable Y follows a normal linear model if its distribution is normal $\mathcal{N}(y|\mu, \sigma^2)$ with mean*

$$\mu = \beta_0 + \beta_1 x_1 + \ldots + \beta_k x_k,$$

$x_1, \ldots, x_k$ *being deterministic variables, and $\beta_0, \ldots, \beta_k$ real parameters.*

This model is usually written as the linear function

$$Y = \beta_0 + \beta_1 x_1 + \ldots + \beta_k x_k + \varepsilon,$$

where ε is a random variable with distribution $\mathcal{N}(\varepsilon|0, \sigma^2)$. The variable Y is called the *response* variable, and the deterministic variables $x_1, \ldots, x_k$ are called *covariates*, or *regressors*, and, as it is assumed that the covariates explain most of the variability of the random variable Y, they are also called *explanatory* variables. The parameters $\beta_0, \ldots, \beta_k$ are called *regression coefficients*, and σ^2 the *residual variance*.

2.6.1   Maximum likelihood estimators

Let $\mathbf{y} = (y_1, \ldots, y_n)^\top$ be a column vector of n independent observations of Y for the covariates $(\mathbf{x}_1, \ldots, \mathbf{x}_n)^\top$, where $\mathbf{x}_i = (x_{i1}, \ldots, x_{ik})$ for $i = 1, \ldots, n$.

Then, observation y_i can be written as

$$y_i = \beta_0 + \sum_{j=1}^{k} \beta_j x_{ij} + \varepsilon_i,$$

where x_{ij} is the j−th value of covariate $\mathbf{x}_i$, and ε_i a random nonobservable error. Thus, $\mathbf{y}$ is written as

$$\mathbf{y} = \mathbf{X}\boldsymbol{\beta} + \boldsymbol{\varepsilon}, \qquad (2.50)$$

where

$$\mathbf{X} = \begin{pmatrix} 1 & x_{11} & \cdots & x_{1k} \\ \vdots & \vdots & \ddots & \vdots \\ 1 & x_{n1} & \cdots & x_{nk} \end{pmatrix}$$

is the *design matrix* of dimensions $n \times (k+1)$, $\boldsymbol{\beta} = (\beta_0, \beta_1, \ldots, \beta_k)^\top$ is the regression coefficient vector, and $\boldsymbol{\epsilon} = (\varepsilon_1, \ldots, \varepsilon_n)^\top$ a random vector of dimension n with the multivariate normal distribution $\mathcal{N}_n(\boldsymbol{\varepsilon}|\mathbf{0}, \sigma^2 \mathbf{I}_n)$, where $\mathbf{I}_n$ is the identity matrix of dimensions $n \times n$. It is assumed that $\mathbf{X}$ is a full rank matrix, that is, $\mathbf{X}^\top \mathbf{X}$ is a nonsingular squared matrix.

The likelihood of $\boldsymbol{\beta}$ and σ for the sample $(\mathbf{y}, \mathbf{X})$ is given by

$$
\begin{aligned}
\ell_{\mathbf{y},\mathbf{X}}(\boldsymbol{\beta}, \sigma) &= \mathcal{N}_n(\mathbf{y}|\mathbf{X}\boldsymbol{\beta}, \sigma^2 \mathbf{I}_n) \qquad (2.51) \\
&= \sigma^{-n} \exp\left\{-\frac{1}{2\sigma^2}(\mathbf{y} - \mathbf{X}\boldsymbol{\beta})^\top (\mathbf{y} - \mathbf{X}\boldsymbol{\beta})\right\}.
\end{aligned}
$$

Maximizing the likelihood function, the MLE of $\boldsymbol{\beta}$ and σ^2 turns out to be

$$\widehat{\boldsymbol{\beta}} = \left(\mathbf{X}^\top \mathbf{X}\right)^{-1} \mathbf{X}^\top \mathbf{y},$$

and

$$\hat{\sigma}^2 = \frac{1}{n}(\mathbf{y} - \mathbf{X}\widehat{\boldsymbol{\beta}})^\top (\mathbf{y} - \mathbf{X}\widehat{\boldsymbol{\beta}}) = \frac{\mathbf{y}^\top (\mathbf{I}_n - \mathbf{H}_n)\mathbf{y}}{n},$$

where $\mathbf{H}_n = \mathbf{X}(\mathbf{X}^\top \mathbf{X})^{-1}\mathbf{X}^\top$ is known as the *hat matrix*.

2.6.2 Bayesian estimators

To derive Bayesian estimators for the parameters we first complete the sampling model with a prior distribution for $\boldsymbol{\beta}$ and σ^2. Conjugate priors for the sampling model (2.51) are given by a multivariate Normal distribution for $\boldsymbol{\beta}$, conditional on σ^2, and an Inverse–Gamma distribution for σ^2. If we write $\pi(\boldsymbol{\beta}, \sigma^2) = \pi(\boldsymbol{\beta}|\sigma^2)\pi(\sigma^2)$ we have the conjugate prior

$$\pi(\boldsymbol{\beta}|\sigma^2) = \mathcal{N}_{k+1}\left(\boldsymbol{\beta}|\mathbf{m}, \sigma^2\mathbf{V}\right),$$
$$\pi(\sigma^2) = \text{IG}(\sigma^2|a, b),$$

where $\mathbf{m} = (m_0, m_1, \ldots, m_k)^\top$ is the prior mean of $\boldsymbol{\beta}$, $\mathbf{V}$ the prior covariance matrix of dimensions $(k+1) \times (k+1)$, and a and b are positive hyperparameters. For the sample $(\mathbf{y}, \mathbf{X})$, the conjugate property of $\pi(\boldsymbol{\beta}, \sigma^2)$ makes it easy to compute the posterior distribution $\pi(\boldsymbol{\beta}, \sigma^2|\mathbf{y}, \mathbf{X})$ which is given by

$$\pi(\boldsymbol{\beta}|\sigma^2, \mathbf{y}, \mathbf{X}) = \mathcal{N}_{k+1}(\boldsymbol{\beta}|\mathbf{m}^\star, \sigma^2\mathbf{V}^\star),$$

and

$$\pi(\sigma^2|\mathbf{y}, \mathbf{X}) = \text{IG}\left(\sigma^2|a^\star, b^\star\right),$$

where

$$\begin{aligned} \mathbf{m}^\star &= \left(\mathbf{X}^\top\mathbf{X} + \mathbf{V}\right)^{-1}\left(\mathbf{X}^\top\mathbf{y} + \mathbf{V}\mathbf{m}\right), \\ \mathbf{V}^\star &= \mathbf{X}^\top\mathbf{X} + \mathbf{V}, \\ a^\star &= a + n, \end{aligned} \tag{2.52}$$

and

$$\begin{aligned} b^\star &= \frac{1}{a^\star}\left[(\mathbf{y} - \mathbf{X}\widehat{\boldsymbol{\beta}})^\top(\mathbf{y} - \mathbf{X}\widehat{\boldsymbol{\beta}}) + ab + (\mathbf{m} - \mathbf{m}^\star)^\top\mathbf{V}(\mathbf{m} - \mathbf{m}^\star)\right. \\ &\qquad \left. + (\widehat{\boldsymbol{\beta}} - \mathbf{m}^\star)^\top\mathbf{X}^\top\mathbf{X}(\widehat{\boldsymbol{\beta}} - \mathbf{m}^\star)\right]. \end{aligned} \tag{2.53}$$

A pair of Bayesian estimators of $\boldsymbol{\beta}$ and σ^2 are the posterior means

$$\mathbb{E}(\boldsymbol{\beta}|\mathbf{y}, \mathbf{X}) = \mathbf{m}^\star,$$

and

$$\mathbb{E}(\sigma^2 | \mathbf{y}, \mathbf{X}) = \frac{b^\star}{a^\star - 1}.$$

We note that the conjugate family needs to elicit $(k^2 + 5k + 8)/2$ hyperparameters, a serious elicitation problem.

A way of mitigating this elicitation problem is to use instead the objective prior

$$\pi(\boldsymbol{\beta}, \sigma) = \frac{c}{\sigma^{k+2}} \, \mathbf{1}_{\mathbb{R}^{k+1} \times \mathbb{R}^+}(\boldsymbol{\beta}, \sigma)$$

where c is an arbitrary positive constant. This prior is improper, and hence the constant c cannot be determined. The posterior distribution of $\boldsymbol{\beta}$ and σ for this prior is given by

$$\pi(\boldsymbol{\beta} | \sigma, \mathbf{y}, \mathbf{X}) = \mathcal{N}_{k+1} \left(\boldsymbol{\beta} | \widehat{\boldsymbol{\beta}}, \sigma^2 \mathbf{X}^\top \mathbf{X} \right),$$

and

$$\pi(\sigma^2 | \mathbf{y}, \mathbf{X}) = \mathrm{IG} \left(\sigma^2 | \frac{n - (k+1)}{2}, \frac{1}{2} (\mathbf{y} - \mathbf{X}\hat{\boldsymbol{\beta}})^\top (\mathbf{y} - \mathbf{X}\hat{\boldsymbol{\beta}}) \right).$$

We note that the posterior mean of $\boldsymbol{\beta}$ coincides with the MLE $\widehat{\boldsymbol{\beta}}$.

The fact that $\pi(\boldsymbol{\beta}, \sigma)$ is an improper prior leaves the marginal of the data

$$m(\mathbf{y}, \mathbf{X}) = c \int_{\mathbb{R}^{k+1} \times \mathbb{R}^+} \frac{1}{\sigma^{k+2}} \mathcal{N}_n(\mathbf{y} | \mathbf{X}\boldsymbol{\beta}, \sigma^2 I_n) \, d\boldsymbol{\beta} \, d\sigma$$

defined up to the arbitrary constant c. This makes this prior unsuitable for model selection in linear models, and in particular for variable selection. The alternative suggested by Berger and Pericchi (1996b) was the use of the intrinsic Bayes factor, a tool that was the seed for the definition of the intrinsic priors.

2.6.3   An outline of variable selection

Information on certain deterministic characteristics such as age, sex, and some others related to a specific disease is typically available from patients receiving a treatment. Linear models are adapted to this situation and thus they are of interest in cost–effectiveness analysis. In this setting arises the subgroup cost–effectiveness analysis (Sculpher and Gafni, 2001; Sculpher, 2008; Espinoza et al., 2014). A subgroup of patients is formed of patients sharing specified values of the covariates. The optimal treatment for a given subgroup of patients now depends on the covariates in the linear models, and hence it might not coincide with the optimal treatment for the whole patient population. Therefore, because the consideration of covariates improves the cost–effectiveness analysis, an important task for finding the optimal treatment for subgroups is that of including in the linear model only those covariates that do have an influence on the cost and the effectiveness of the treatments.

The statistical problem of selecting the influential covariates from a potential set of them is an old and central problem in the statistical literature on linear models, and it is known as the *variable selection problem*. This problem has been given several alternative frequentist and Bayesian solutions over the years. The solution proposed by the objective Bayesian methodology will be presented in Chapter 6 where the subgroup cost–effectiveness analysis is developed.

3

Statistical decision theory

3.1 Introduction

In this chapter we provide the elements, fundamental concepts, and
basic results of the statistical decision theory. These results will be used
in the next chapters of the book. The statistical decision theory starts
with a space of alternative decisions $\mathcal{D} = \{d\}$ such that each decision
d has associated with it a reward so that we have the reward space
$\mathcal{R} = \{r_d : d \in \mathcal{D}\}$, and the problem consists of choosing the decision
having the *preferred* reward in $\mathcal{R}$, which is called the *optimal* decision.

In most of real applications of decision theory, the reward of a decision is not deterministic but an observable random variable r which
is dictated by the decision problem. Thus, as r is a random variable,
it has to be described by a probability distribution, and the reward of
the decision d becomes the probability distribution $P_d(r)$. Therefore,
in the space of rewards $\mathcal{P} = \{P_d(r) : d \in \mathcal{D}\}$, an ordering of preference
has to be given in order to choose an optimal decision.

It is really hard to figure out an ordering of preference between
probability distributions and it is not surprising that strong conditions
have to be imposed to achieve such an ordering. However, we cannot
escape from it.

A brief summary on how the ordering is defined is as follows. For
an arbitrary but fixed ordering of the distributions in the space $\mathcal{P}$,
it is shown that there exists a utility function $U(r)$ of the random
reward r valued in the real line that satisfies the *faithful* condition
for that ordering, that is, $P_{d'}(r)$ is preferred to $P_{d''}(r)$ if and only if

$\mathbb{E}_{P_{d'}}(U) > \mathbb{E}_{P_{d''}}(U)$, where $\mathbb{E}_{P_d}(U)$ denotes the expectation of $U(r)$ with respect to $P_d(r)$, that is,

$$\mathbb{E}_{P_d}(U) = \int_{r \in \mathcal{R}} U(r) \, dP_d(r), \quad d \in \mathcal{D}.$$

To obtain this result, the ordering of preference of the rewards must satisfy a set of four *rational behavior* axioms. These axioms are discussed in the light of the consequences derived from them.

A *rational* decision maker is one who uses for decision making a utility function. As a consequence, different *rational* decision makers might use different utility functions depending whether they are either *adverse to the risk, lover risk*, or hold an *intermediate position*.

The organization of the chapter is as follows. In Section 3.2 the elements of a decision problem and the ordering between reward distributions are formulated. In Section 3.3 it is assumed that the reward probability distributions in $\mathcal{P}$ are completely specified, and the existence of the faithful utility function is presented. In Section 3.4 and 3.5 it is assumed that the reward probability distributions in $\mathcal{P}$ are not completely specified but they are parametric classes of probability distributions $P_d(r|\theta)$, $d \in \mathcal{D}$, where θ is an unknown parameter. In this case, either a strategy to eliminate the unknown parameter θ is assumed, for instance a minimax or a Bayesian strategy, or it is eliminated by assuming that a sample $(r_1, \ldots, r_n)$ from $P_d(r|\theta)$ is available. The reason to impose the latter assumption is that now the decision–making procedure in the presence of sampling information can employ tools of statistical inference to manage the uncertainty in θ. In this case, both the frequentist and the Bayesian statistical inference approaches are considered.

3.2   Elements of a decision problem

A decision problem arises when one has a space of decisions $\mathcal{D} = \{d_1, \ldots, d_k\}$, and the choice of decision d_i yields a reward r_i for $i = 1, \ldots, k$, that is not necessarily a monetary real quantity.

Although the decision space $\mathcal{D}$ could be continuous, we keep in mind that the decision problem we are interested in is that of choosing a medical treatment among a finite collection of treatments whose rewards are defined in terms of their cost and effectiveness. Thus, we restrict ourselves to finite decision spaces.

The problem consists of choosing a decision in $\mathcal{D}$ having the preferred reward. In this section we introduce the elements of a statistical decision problem, the notion of preference between rewards, and the standard methodology for making optimal decisions.

3.2.1   Ordering rewards

We assume that for any $i = 1, \ldots, k$, the reward of the decision d_i is a point r_i that belongs to the reward space $\mathcal{R}$, which is usually either the space $\mathbb{R}^p$ for $p \geq 1$, where $\mathbb{R}$ is the real line, or a discrete subset of $\mathbb{R}^p$. Because of the need to establish a preference between rewards, we assume that there exists a binary relationship $\preceq$ in $\mathcal{R}$ such that any pair of rewards r_1 and r_2 in $\mathcal{R}$ can be compared.

We shall write $r_1 \prec r_2$ if r_2 is preferred to r_1, $r_1 \preceq r_2$ if r_1 is not preferred to r_2, and $r_1 \sim r_2$ if r_1 and r_2 are equivalent. We also assume that the binary relationship $\preceq$ provides a complete ordering in the space $\mathcal{R}$. This means that the binary relationship $\preceq$ satisfies the following three properties:

I.   For any two rewards r_1 and r_2, only one of the relationships

$$r_1 \prec r_2, \quad r_2 \prec r_1, \quad r_1 \sim r_2,$$

holds.

II.   If r_1 and r_2 are such that $r_1 \preceq r_2$ and $r_2 \preceq r_1$, then $r_1 \sim r_2$.

III.   If r_1, r_2 and r_3 are such that $r_1 \preceq r_2$ and $r_2 \preceq r_3$, then $r_1 \preceq r_3$.

The existence of the binary relationship $\preceq$ in $\mathcal{R}$ satisfying I–III implies that $(\mathcal{R}, \preceq)$ is an *ordered space*.

Given that the choice of a decision d_i provides the reward r_i for any $i = 1, \ldots, k$, the optimal decision is clearly the one that corresponds to the most preferred reward. That is, d_j is the optimal decision with respect to the binary relationship $\preceq$ if $r_i \preceq r_j$ for any $i = 1, \ldots, k$.

3.2.2   Lotteries

For most practical decision problems, the reward of a decision d is rarely a fixed reward r, but typically r is a random reward that follows a given probability distribution $P_d(r)$ for $r \in \mathcal{R}$. Thus, we will assume that any decision $d_i \in \mathcal{D}$ has associated a probability distribution $P_i(r)$ on $\mathcal{R}$. This reward distribution is also called a *lottery*.

In what follows, by $(\mathcal{R}, \mathcal{A}, P)$ we denote a probability space on the set of rewards $\mathcal{R}$, where $\mathcal{A}$ is a class of subsets of $\mathcal{R}$, and P a probability distribution. The class $\mathcal{A}$ is a $\sigma-$field, which is a class that is closed with respect to countable unions and complementations of sets in $\mathcal{A}$.

For instance, when we choose a treatment T_i from the set of alternative treatments $\{T_1, \ldots, T_k\}$ in a cost–effectiveness analysis, the reward we obtain is a probability distribution of the effectiveness and cost of the treatment. A frequent case is the one where the effectiveness of treatment T_i is independent of the cost and it follows a Bernoulli distribution with probability of success θ_i, that is, $\mathrm{Be}(e|\theta_i) = \theta_i^e (1 - \theta_i)^{1-e}$ for $e = 0, 1$ and $0 \leq \theta_i \leq 1$. Assuming that the probability distribution of the cost of the treatment is lognormal $\Lambda(c|\mu_i, \sigma_i^2)$, the lottery we obtain when choosing treatment T_i is given by $P_i(c, e|\mu_i, \sigma_i, \theta_i) = \Lambda(c|\mu_i, \sigma_i^2)\mathrm{Be}(e|\theta_i)$, a probability distribution in the $\sigma-$field $\mathcal{A} = \mathcal{B}^+ \times \{0, 1\}$, where $\mathcal{B}^+$ is the Borel $\sigma-$field in $\mathbb{R}^+$. If we assume that the parameters μ_i, σ_i and θ_i are known, the lottery is fully determined, and otherwise a strategy for eliminating them from the lottery is needed.

In this section we assume that the lottery of any decision is fully determined, and the situation where lotteries are not fully determined is discussed in Sections 3.3 and 3.4.

In what follows, the space of rewards $\mathcal{P}$ is the space of all distributions on $\mathcal{R}$ so that $\mathcal{P}$ contains the lotteries $P_i(r)$ for $i = 1, \ldots, k$, and also any reward $r \in \mathcal{R}$ which is considered as the Dirac's distribution $\delta_r(r')$ on point r, $r' \in \mathcal{R}$.

On the set of lotteries $\mathcal{P}$ we assume that there exists a binary relationship of preference between them that we denote as $\preceq^*$. This means that for two arbitrary distributions P_1 and P_2 in $\mathcal{P}$ we can write $P_1 \prec^* P_2$ if P_2 is preferred to P_1, $P_1 \preceq^* P_2$ if P_1 is not preferred to P_2, and $P_1 \sim^* P_2$ if P_1 and P_2 are equivalent. We note that the binary relationship $\preceq^*$ coincides with $\preceq$ for the degenerated rewards $\delta_r(r') \equiv r$ in $\mathcal{R}$.

We assume that the binary relationship $\preceq^*$ satisfies the following three properties:

I*. For any two rewards P_1 and P_2, only one of the following three relationships

$$P_1 \prec^* P_2, \quad P_2 \prec^* P_1, \quad P_1 \sim^* P_2,$$

holds.

II*. If P_1 and P_2 are such that $P_1 \preceq^* P_2$ and $P_2 \preceq^* P_1$, then $P_1 \sim^* P_2$.

III*. If P_1, P_2, and P_3 are such that $P_1 \preceq^* P_2$ and $P_2 \preceq^* P_3$, then $P_1 \preceq^* P_3$.

The existence of the binary relationship $\preceq^*$ in $\mathcal{P}$ satisfying I*–III* implies that $(\mathcal{P}, \preceq^*)$ is an ordered space.

One may think that the set of assumptions I*–III* are a very demanding set of assumptions. This is particularly true for assumption I*, which is really a strong assumption, although strong conditions are necessary for rational decision making.

3.2.3 The utility function

Let $(\mathcal{R}, \mathcal{A}, P)$ be a probability space on the set of rewards $\mathcal{R}$. Given an $\mathcal{A}$-measurable real function $U(r)$ defined on the reward space $\mathcal{R}$, the expectation of $U(r)$ with respect to $P(r)$ is the integral of $U(r)$ with respect to the probability distribution $P(r)$, and it is denoted as

$$\mathbb{E}_P(U) = \int_{\mathcal{R}} U(r) \, dP(r). \tag{3.1}$$

We are interested in a special class of functions $U(r)$ defined on the probability space $(\mathcal{R}, \mathcal{A}, P)$, for P belonging to the ordered set of distributions $(\mathcal{P}, \preceq^*)$. The members of this special class of functions are called *utility functions*, and are defined as follows.

Definition 3.1. *A real–valued function $U(r) : \mathcal{R} \longrightarrow \mathbb{R}$ is a utility function if for any pair of distributions $P_1, P_2 \in \mathcal{P}$ such that $\mathbb{E}_{P_1}(U) < \infty$ and $\mathbb{E}_{P_2}(U) < \infty$, the equivalence*

$$P_1 \preceq^* P_2 \Longleftrightarrow \mathbb{E}_{P_1}(U) \leq \mathbb{E}_{P_2}(U) \tag{3.2}$$

holds.

The expectation $\mathbb{E}_P(U)$ is called the *utility of P*, and for any degenerate reward $r \in \mathcal{R}$ the real number $U(r)$ is the utility of r.

The existence of a utility function $U(r)$ means that the ordering of the real numbers $\{ \, \mathbb{E}_P(U), \ P \in \mathcal{P} \, \}$ is equivalent to the ordering of the lotteries in the space $(\mathcal{P}, \preceq^*)$. Thus, given a decision problem, it is of utmost importance to establish conditions under which there exists a utility function $U(r)$, a point that we discuss in the next section.

As a consequence of the existence of a utility function $U(r)$ we note that

$$r_1 \preceq r_2 \Longleftrightarrow U(r_1) \leq U(r_2)$$

since the degenerate distributions $\delta_{r_1}(r)$ and $\delta_{r_2}(r)$ are in $\mathcal{P}$.

We also note that for any function $W(r) = aU(r) + b$, where $a > 0$ is an scale parameter and b a location parameter, we have that

$$\mathbb{E}_{P_1}(W) \leq \mathbb{E}_{P_2}(W) \Longleftrightarrow \mathbb{E}_{P_1}(U) \leq \mathbb{E}_{P_2}(U).$$

This equivalence means that any linear combination of a utility function is also a utility function. Consequently, a utility function is defined up to a scale and location parameters.

3.3 Axioms for the existence of the utility function

The purpose of the utility theory is to establish a set of conditions on the binary relationship $\preceq^*$ in the space of lotteries $\mathcal{P}$ such that a unique utility function $U(r)$ can be constructed. The first set of axioms on $\preceq^*$ to guarantee the existence of a utility function were given by von Neumann and Morgenstern (1944), and the axioms we give below have been taken from the book by DeGroot (1970). We refer the reader to this book for a more detailed development of the utility theory.

Axiom 1. *Let P_1, P_2 and P be three distributions in $\mathcal{P}$. Let $0 < \alpha < 1$. Then,*

$$P_1 \prec^* P_2 \iff \alpha P_1 + (1 - \alpha)P \prec^* \alpha P_2 + (1 - \alpha)P.$$

The meaning of Axiom 1 is that a common part of a mixture of distributions does not alter the sign of the binary relationship. An important consequence of this axiom is that for two rewards such that $r_1 \prec r_2$ and any $\alpha \in (0, 1)$ we have that $r_1 \prec \alpha r_2 + (1 - \alpha)r_1 \prec r_2$. We note that the mixture $\alpha r_2 + (1 - \alpha)r_1$ is a lottery whose meaning is that reward r_2 is obtained with probability α and r_1 with probability $1 - \alpha$.

Axiom 2. *Let P_1, P_2 and P be three distributions in $\mathcal{P}$ such that $P_1 \prec^* P \prec^* P_2$. Then, there exist two numbers $\alpha, \beta \in [0, 1]$ such that*

$$P \prec^* \alpha P_1 + (1 - \alpha)P_2 \quad and \quad P \succ^* \beta P_1 + (1 - \beta)P_2.$$

An important consequence of this axiom is that if the preference relationship $r_1 \prec r \prec r_2$ holds, it is not enough to avoid that $r \prec^* \alpha r_2 + (1 - \alpha)r_1$ for a value $\alpha \in (0, 1)$. Following DeGroot (1970) this means that there is no "heaven" in $\mathcal{R}$. Similarly, even when r_1 is less preferred to r it is not enough to avoid that $r \succ^* \beta r_2 + (1 - \beta)r_1$ for a value $\beta \in (0, 1)$. That is, there is no "hell" in $\mathcal{R}$.

A consequence of Axioms 1 and 2 is that if $r_1 \preceq r \preceq r_2$, there exists a unique number $\gamma(r) \in [0,1]$ such that

$$r \sim^* \gamma(r)r_2 + (1 - \gamma(r))r_1 \,. \tag{3.3}$$

This result allows us to define a function $U(r)$ as follows.

Step 1. We first consider two reference rewards s_0 and t_0 such that $s_0 \prec t_0$, and then proceed defining,

Step 2.

a)   For each r such that $s_0 \preceq r \preceq t_0$,

$$U(r) = \alpha,$$

where α is the unique number such that $r \sim^* \alpha t_0 + (1 - \alpha)s_0$. We note that $U(s_0) = 0$ and $U(t_0) = 1$.

b)   For each r such that $r \prec s_0 \prec t_0$,

$$U(r) = -\frac{\alpha}{1 - \alpha},$$

where α is the unique number such that $s_0 \sim^* \alpha t_0 + (1 - \alpha)r$.

c)   For each r such that $s_0 \prec t_0 \prec r$,

$$U(r) = \frac{1}{\alpha},$$

where α is the unique number such that $t_0 \sim^* \alpha r + (1 - \alpha)s_0$.

A function $U(r)$ thus defined can be shown to be "linear," that is, for any three rewards $r_1, r_2, r_3 \in \mathcal{R}$ such that $r_2 \sim^* \gamma r_3 + (1 - \gamma)r_1$ it follows that

$$U(r_2) = \gamma U(r_3) + (1 - \gamma)U(r_1).$$

Axiom 3. *For any three rewards $r_1, r_2, r_3 \in \mathcal{R}$ and any $\alpha, \beta \in [0,1]$, the set of rewards $A(\alpha, \beta) = \{ r : \alpha r + (1 - \alpha)r_1 \preceq^* \beta r_2 + (1 - \beta)r_3 \}$ is a measurable set, that is,*

$$A(\alpha, \beta) \in \mathcal{A}.$$

The consequence of this measurability Axiom 3 is that the utility function $U(r)$ defined above is an $\mathcal{A}$-measurable function, so that the expectation with respect to the distribution P defined on $\mathcal{A}$ is well defined.

So far, $U(r)$ has been defined with respect to the reference rewards s_0 and t_0, with $s_0 \prec t_0$, and hence $U(s_0) = 0$ and $U(t_0) = 1$. To define the utility with respect to an arbitrary pair of rewards r_1 and r_2 such that $r_1 \prec r_2$ we can proceed as follows. First, we note that for any $r \in [r_1, r_2] = \{r : r_1 \preceq r \preceq r_2\}$ there exits a unique number $\gamma(r)$ such that

$$r \sim^* \gamma(r) r_2 + (1 - \gamma(r)) r_1.$$

Further, the number $\gamma(r)$ can be written in terms of $U(r)$ defined above. Indeed, from $U(r) = \gamma(r) U(r_2) + (1 - \gamma(r)) U(r_1)$, we have that

$$\gamma(r) = \frac{U(r) - U(r_1)}{U(r_2) - U(r_1)}.$$

This suggests that any P would be equivalent to $\beta r_2 + (1 - \beta) r_1$ for an appropriate β. This β would be

$$\beta = \int_{[r_1, r_2]} \gamma(r) \, dP(r).$$

This is formalized in the following Axiom 4.

Axiom 4. *Let P be a distribution such that $P[r_1, r_2] = 1$. Then, we accept that*

$$P \sim^* \beta r_2 + (1 - \beta) r_1,$$

where $\beta = \int_{[r_1, r_2]} \gamma(r) \, dP(r)$.

The main consequence of Axiom 4 is that we can write P in terms of r_1 and r_2 as

$$P \sim^* \frac{\mathbb{E}_P(U) - U(r_1)}{U(r_2) - U(r_1)} r_2 + \frac{U(r_2) - \mathbb{E}_P(U)}{U(r_2) - U(r_1)} r_1.$$

Under Axioms 1 to 4 it can be proved that for two distributions P_1 and P_2 in $\mathcal{P}$ such that $\mathbb{E}_{P_1}(U) < \infty$ and $\mathbb{E}_{P_2}(U) < \infty$, we have that

$$P_1 \preceq^* P_2 \iff \mathbb{E}_{P_1}(U) \leq \mathbb{E}_{P_2}(U).$$

Since the utility function is defined except for a location and scale parameter, without loss of generality we can assume that the utility function is valued in the positive part of the real line, that is $U(r) : \mathcal{R} \longrightarrow \mathbb{R}^{+}$.

Another consequence of the above developments is that if one has a utility function $U(r)$ for a decision problem, the optimal decision d is the one having the reward P that maximizes the expected utility, that is

$$\mathbb{E}_P(U) = \sup_{Q \in \mathcal{P}} \mathbb{E}_Q(U). \tag{3.4}$$

3.4   Criticisms of the utility function

Despite numerous criticisms, the utility theory framework has been accepted by the vast majority of decision makers. An acute criticism of the existence of a utility function is presented in the following example given by Allais (1953).

The example assumes a monetary reward space with three elements $\mathcal{R} = \{r_1, r_2, r_3\}$, where $r_1 = \$2,500,000$, $r_2 = \$500,000$ and $r_3 = \$0$, and a utility function $U(r)$ such that

$$U(\$2,500,000) > U(\$500,000) > U(\$0).$$

Let us consider the two lotteries, one given by

$$P_1(r) = \begin{cases} 0, & \text{if } r = r_1, \\ 1, & \text{if } r = r_2, \\ 0, & \text{if } r = r_3, \end{cases}$$

for which we get $\$500,000$ with probability 1, and a second given by

$$P_2(r) = \begin{cases} 0.10, & \text{if } r = r_1, \\ 0.89, & \text{if } r = r_2, \\ 0.01, & \text{if } r = r_3, \end{cases}$$

for which we get \$2,500,000 with probability 0.10, \$500,000 with probability 0.89, and \$0 with probability 0.01. The utility of the lotteries P_1 and P_2 are

$$\mathbb{E}_{P_1}(U) = U(\$500,000),$$

and

$$\mathbb{E}_{P_2}(U) = 0.01 \, U(\$0) + 0.10 \, U(\$2,500,000) + 0.89 \, U(\$500,000).$$

On the other hand, let us consider two more lotteries

$$P_3(r) = \begin{cases} 0.00, & \text{if } r = r_1, \\ 0.11, & \text{if } r = r_2, \\ 0.89, & \text{if } r = r_3, \end{cases}$$

and

$$P_4(r) = \begin{cases} 0.10, & \text{if } r = r_1, \\ 0.00, & \text{if } r = r_2, \cdot \\ 0.90, & \text{if } r = r_3. \end{cases}$$

The utility of P_3 and P_4 are

$$\mathbb{E}_{P_3}(U) = 0.89 \, U(\$0) + 0.11 \, U(\$500,000),$$

$$\mathbb{E}_{P_4}(U) = 0.90 \, U(\$0) + 0.10 \, U(\$2,500,000).$$

We note that $\mathbb{E}_{P_2}(U) - \mathbb{E}_{P_1}(U) = \mathbb{E}_{P_4}(U) - \mathbb{E}_{P_3}(U)$, and hence it is satisfied that

$$\mathbb{E}_{P_2}(U) > \mathbb{E}_{P_1}(U) \iff \mathbb{E}_{P_4}(U) > \mathbb{E}_{P_3}(U).$$

From this equivalence, it follows from the utility theory that P_2 is preferred to P_1 if and only if P_4 is preferred to P_3. One can argue that under P_1 a gain of \$500,000 is obtained with probability 1, and that the reward P_2 has a positive probability of gain \$0. Then it might be that for many people $P_1 \succ^* P_2$. This implies that for these people, P_3 should be preferred to P_4.

However, one might think that P_4 has only a slightly greater probability of getting \$0 than P_3, whereas the probability of gaining \$2,500,000 is much larger under P_4 than under P_3, so that for many people, P_4 is preferred to P_3, and at the same time they prefer P_1 to P_2.

This behavior contradicts the ordering of the lotteries given by the utility function $U(r)$, and suggests that the ordering of the lotteries might depend on something more than the utility $U(r)$ of the degenerate rewards $r \in \mathcal{R}$. Further discussion of this contradiction can be found in Savage (1954) and DeGroot (1970).

3.5   Lotteries that depend on a parameter

So far the lotteries $P_i(r)$, $i = 1, \ldots, k$ were fully determined. However, as we mentioned at the beginning of Section 3.1, in most decision problems, the reward we obtain when making the decision $d_i \in \mathcal{D}$ is not fully determined but it depends on a nonobservable parameter $\theta_i \in \Theta$. Thus, the probability distribution $P_i(r)$ now becomes $P_i(r|\theta_i)$.

In this setting, the expectation of the utility function $U(r)$ with respect to the reward $P_i(r|\theta_i)$ of the decision d_i is defined up to the unknown parameter θ_i, that is, the utility of the lottery P_i becomes

$$\varphi(d_i|\theta_i) = \int_{\mathcal{R}} U(r)\, dP_i(r|\theta_i). \tag{3.5}$$

As a consequence, two decisions d_i and d_j are not comparable because the utility of their rewards are functions of θ_i and θ_j, respectively, unless the utility of one lottery is uniformly greater than the utility of the other lottery. For instance, if $\varphi(d_j|\theta_j) \geq \varphi(d_i|\theta_i)$ for any θ_i and θ_j in Θ, then d_j is preferred to d_i. In this case it is said that d_j *dominates* d_i, and d_i is said to be an *inadmissible decision*. The subset of inadmissible decisions must be suppressed from the decision space.

The class with no inadmissible decision is called the *admissible class of decisions*. Consequently, for any two decisions d_i and d_j in the admissible class the utility of their rewards $P_i(r|\theta_i)$ and $P_j(r|\theta_j)$ are not comparable. In the following sections, two strategies to deal with this situation are presented.

3.5.1 The minimax strategy

It is usual in statistical decision theory to consider the loss function $\mathcal{L}(r)$, which is defined as minus the utility function, that is

$$\mathcal{L}(r) = -U(r), \ r \in \mathcal{R}.$$

The minimax strategy consists of selecting a decision, called a *minimax decision*, which is obtained as follows. The expectation of the loss function $\mathcal{L}(r)$ with respect to the lottery $P_i(r|\theta_i)$ is called the *risk of the decision d_i*, that is

$$\rho(d_i|\theta_i) = \int_{r \subset \mathcal{R}} \mathcal{L}(r) \, dP_i(r|\theta_i).$$

The idea is to first maximize the risk $\rho(d_i|\theta_i)$ with respect to $\theta_i \in \Theta$, that is

$$\hat{\rho}(d_i) = \sup_{\theta_i \in \Theta} \rho(d_i|\theta_i),$$

for $i = 1, \ldots, k$. Then, the maximized risks of the possible decisions $\{\hat{\rho}(d_i), i = 1, \ldots, k\}$ are minimized. This strategy means that we are being "pessimistic" with respect to the unknown parameters $\{\theta_i, i = 1, \ldots, k\}$ as the worst possible parameter values are being accepted.

Thus, the minimax decision in the class $\{d_1, \ldots, d_k\}$ is d_j if the equality

$$\hat{\rho}(d_j) = \min_{i=1,\ldots,k} \hat{\rho}(d_i)$$

holds.

3.5.2 The Bayesian strategy

Given the set of lotteries $\{P_i(r|\theta_i), \theta_i \in \Theta, i = 1, \ldots, k\}$, the Bayesian strategy first completes the lotteries by assigning a prior probability distribution $\pi_i(\theta_i)$ to parameter θ_i. Then, the *Bayesian risk of decision*

d_i is defined as the expectation with respect to $\pi_i(\theta_i)$ of the risk function $\rho(d_i|\theta_i)$, that is,

$$\rho(d_i, \pi_i) = \int_\Theta \rho(d_i|\theta_i)\pi_i(\theta_i) \, d\theta_i. \tag{3.6}$$

This can be also written as

$$\rho(d_i, \pi_i) = \int_{\theta_i \in \Theta} \int_{r \in \mathcal{R}} \mathcal{L}(r) \, dP_i(r|\theta_i) \, \pi_i(\theta_i) \, d\theta_i. \tag{3.7}$$

The *optimal Bayesian decision* for the priors $\{\pi_i(\theta_i), i = 1, \dots, k\}$, is the decision that minimizes the risks $\{\rho(d_i, \pi_i), \ i = 1, \dots, k\}$. Thus, d_j is an optimal Bayesian decision if the equation

$$\rho(d_j, \pi_j) = \min_{i=1,\dots,k} \rho(d_i, \pi_i)$$

holds.

3.5.3   Comparison

The risk of the Bayesian optimal decision is smaller than or equal to the risk of the minimax decision. Indeed, for decision d_i and any prior distribution $\pi_i(\theta_i)$ we have that its Bayesian risk satisfies

$$
\begin{aligned}
\rho(d_i, \pi_i) &= \int_\Theta \rho(d_i|\theta_i)\pi_i(\theta_i) \, d\theta_i \leq \sup_{\theta_i \in \Theta} \rho(d_i|\theta_i) \int_\Theta \pi_i(\theta_i) \, d\theta_i \\
&= \sup_{\theta_i \in \Theta} \rho(d_i|\theta_i) = \hat{\rho}(d_i),
\end{aligned}
$$

for $i = 1, \dots, k$. Thus, it follows that

$$\min_{i=1,\dots,k} \rho(d_i, \pi_i) \leq \min_{i=1,\dots,k} \hat{\rho}(d_i).$$

This means that the risk of the minimax strategy is unnecessarily large.

We mention in passing the interesting result that the class of admissible decisions are either Bayesian decisions or the limit of sequences of Bayesian decisions. This result was proved by Wald (1971) and confirms the assertion that the Bayesian is preferred to the minimax strategy.

3.6 Optimal decisions in the presence of sampling information

In the preceding sections we did not use any empirical information on the reward distribution $\{P_i(r|\theta_i),\ \theta_i \in \Theta\}, i = 1, \ldots, k$. However, in decision problems there are typically available samples from these distributions. For instance, when we consider the problem of choosing a treatment T_i among a set of alternative treatments $\{T_1, \ldots, T_k\}$, it is usual that we have samples of the effectiveness and cost from patients under the treatments. This information can be employed for eliminating the unknown parameters $\{\theta_1, \ldots, \theta_k\}$ from the lotteries.

Therefore, we consider the situation in which there is an experiment providing sampling information $\mathbf{r}_i = (r_{i1}, \ldots, r_{in_i})$ of size n_i from the distribution $P_i(r|\theta_i), i = 1, \ldots, k$. Then, the likelihood of the parameter θ_i for the sample $\mathbf{r}_i$ is given by

$$P_i(\mathbf{r}_i|\theta_i) = \prod_{j=1}^{n_i} P_i(r_{ij}|\theta_i). \qquad (3.8)$$

At this point we have two alternative procedures for incorporating the sample information into the decision making that we briefly describe in the next two subsections.

3.6.1 The frequentist procedure

This procedure is based on the maximum likelihood estimator $\hat{\theta}_i = \hat{\theta}_i(\mathbf{r}_i)$ of θ_i, given by

$$\hat{\theta}_i = \arg \sup_{\theta_i \in \Theta_i} P(\mathbf{r}_i|\theta_i),$$

for $i = 1, \ldots, k$. The frequentist procedure simply plugs in $\hat{\theta}_i$ in the original set of lotteries $\mathcal{P} = \{P_i(r|\theta_i),\ i = 1, \ldots, k\}$ that is replaced by the estimated set of lotteries

$$\mathcal{P}' = \{P_i(r|\hat{\theta}_i),\ i = 1, \ldots, k\}.$$

We note that $P_i(r|\hat{\theta}_i)$ is the predictive lottery for the data $\mathbf{r}_i$, introduced in Section 2.4. The expected utility of $P_i(r|\hat{\theta}_i)$ is given by

$$\varphi(d_i|\hat{\theta}_i) = \int_{\mathcal{R}} U(r)\, dP_i(r|\hat{\theta}_i).$$

In this setting, decision d_j is optimal if the equality

$$\varphi(d_j|\hat{\theta}_j) = \max_{i=1,\dots,k} \varphi(d_i|\hat{\theta}_i)$$

holds. Thus, the optimal decision d_j depends on the data $\{\mathbf{r}_i, i = 1,\dots,k\}$ through the maximum likelihood estimators $\{\hat{\theta}_i(\mathbf{r}_i),\ i = 1,\dots,k\}$.

3.6.2   The Bayesian procedure

Let $\pi_i(\theta_i)$ be a prior distribution of parameter θ_i of the lottery $P_i(r|\theta_i)$, $i = 1,\dots,k$. The Bayesian procedure for incorporating the sampling information $\{\mathbf{r}_i, i = 1,\dots,k\}$ into the decision problem is based on the posterior distribution of θ_i

$$\pi_i(\theta_i|\mathbf{r}_i) = \frac{P_i(\mathbf{r}_i|\theta_i)\pi_i(\theta_i)}{\int_{\Theta} P_i(\mathbf{r}_i|\theta_i)\pi_i(\theta_i)d\theta_i},$$

where $P_i(\mathbf{r}_i|\theta_i)$ is the likelihood of θ_i in (3.8).

We now proceed to integrate out θ_i from the reward distribution $P_i(r|\theta_i)$ using the posterior distribution of θ_i, and lottery $P_i(r|\theta_i)$ becomes

$$P_i(r|\mathbf{r}_i) = \int_{\Theta} P_i(r|\theta_i)\pi_i(\theta_i|\mathbf{r}_i)d\theta_i.$$

This reward is simply the predictive lottery, Section 2.4, conditional on $\mathbf{r}_i$, of the Bayesian lottery $(P_i(r|\theta_i), \pi_i(\theta_i))$.

Then, instead of considering the undetermined lotteries $\mathcal{P} = \{P_i(r|\theta_i),\ i = 1,\dots,k\}$, we consider the predictive lotteries, conditional on $\mathbf{r}_i$, given by

$$\mathcal{P}'' = \{P_i(r|\mathbf{r}_i), i = 1,\dots,k\}.$$

The *Bayesian expected utility of decision d_i*, conditional on the sample $\mathbf{r}_i$, is then given by

$$\varphi(d_i|\pi_i, \mathbf{r}_i) = \int_{\mathcal{R}} U(r) \, dP_i(r|\mathbf{r}_i),$$

which depends on the prior $\pi_i(\theta_i)$ and the sample $\mathbf{r}_i$.

The *Bayesian optimal decision*, conditional on the sample $\mathbf{r}_i$, is the one having maximum expected utility, that is, d_j is optimal if

$$\varphi(d_j|\pi_i, \mathbf{r}_j) = \max_{i=1,\ldots,k} \varphi(d_i|\pi_i, \mathbf{r}_i). \tag{3.9}$$

We keep in mind that the Bayesian optimal decision depends on the prior distributions $\{\pi_i(\theta_i), \ i = 1, \ldots, k\}$ and the samples $\{\mathbf{r}_i, \ i = 1, \ldots, k\}$.

4

Cost–effectiveness analysis: Optimal treatments

4.1 Introduction

Cost–effectiveness analysis is a statistical decision problem in which, for a given disease, there is a finite set of alternative medical treatments $\{T_1, \ldots, T_k\}$, $k \geq 2$, and the problem consists of choosing the optimal treatment based on its cost and effectiveness. In this chapter we introduce the elements of this decision problem, the utility functions commonly used, and the procedure for characterizing the optimal treatment.

The elements of this decision problem are i) the decision space $\mathcal{D} = \{d_1, \ldots, d_k\}$, where d_i is the decision of choosing treatment T_i, and ii) the reward space $\{P_1(c, e), \ldots, P_k(c, e)\}$, where $P_i(c, e)$ is the reward of the decision d_i, a probability distribution (lottery) of the cost and the effectiveness of treatment T_i, $i = 1, \ldots, k$. The cost is a continuous positive real random variable and the effectiveness is either a continuous or discrete random variable depending on the specific disease. Most of the frequently used measures of effectiveness of medical treatments were discussed in Chapter 1.

If we assume the existence of a utility function $U(c, e)$ for $(c, e) \in \mathcal{R}$, the reward space is the set of distributions

$$
\mathcal{P} = \left\{ P_i(c, e) : \mathbb{E}_{P_i}(U) = \int_{\mathcal{R}} U(c, e) \, dP_i(c, e) < \infty, \ i = 1, \ldots, k \right\}.
$$

From the central result of the decision theory (Chapter 3) it follows that the utility function $U(c, e)$ induces an ordering $\preceq^*$ of preference between the distributions in $\mathcal{P}$ such that

$$P_i(c, e) \preceq^* P_j(c, e) \iff \mathbb{E}_{P_i}(U) \leq \mathbb{E}_{P_j}(U).$$

As a consequence, the decision d_j is an optimal decision in $\mathcal{D}$ for the utility function $U(c, e)$ if the expectation of $U(c, e)$ with respect to $P_j(c, e)$, which is called *the utility of the distribution* $P_j(c, e)$, is the largest one, that is, if the equation

$$\mathbb{E}_{P_j}(U) = \max_{i=1,\dots,k} \mathbb{E}_{P_i}(U)$$

holds. Therefore, the optimal treatment strongly depends on the utility function $U(c, e)$ we adopt for the decision problem.

The utility functions we present in this chapter are those typically used in cost–effectiveness analysis and they are defined in two steps.

In a first step the bidimensional space of cost and effectiveness (c, e) is reduced to the one–dimensional space $\mathcal{Z}$, a subset of the real line $\mathbb{R}$. $\mathcal{Z}$ is the space of the *net benefit* z of (c, e), a notion that we present in Section 4.2. The price we pay for the dimension reduction is the introduction of a new nonnegative real parameter R in the definition of z which is tied to the decision maker.

In a second step the utility function $U(z|R)$ of the net benefit, conditional on R, is defined. Two utility functions of z along with the procedure for computing optimal decisions are given in Section 4.3. The frequent case of two treatments in which one treatment is the *status quo* and the other a new treatment will be considered in Section 4.4. In this situation we argue that the transition costs should be incorporated into the analysis, and this yields a slight modification of the decision rule.

For clarity in the presentation, all the developments in Sections 4.3 and 4.4 are carried out for the case where the reward distributions $\{P_i(c, e), \ i = 1, \dots, k\}$ are completely specified.

In Section 4.5 we extend the decision–making process to the more realistic case where the reward distribution $P_i(c, e)$ depends on an unknown parameter $\theta_i \in \Theta$, so that the distribution is now written as $P_i(c, e | \theta_i)$ for $i = 1, \ldots, k$. This implies that the optimal decision now depends on this unknown parameter that has to be eliminated from these distributions. The elimination is carried out using samples of the cost and effectiveness. There are two ways of doing that, and we present both the Bayesian and the frequentist (see Section 2.4 in Chapter 2). The resulting distributions are the *Bayesian and frequentist predictive distributions*. We will see that the Bayesian and frequentist optimal treatments do not necessarily coincide, a consequence of the different ways the sampling information is taken into account by these predictive distributions.

Section 4.6 is devoted to illustrating the frequentist and Bayesian predictive distributions on statistical models of frequent use in cost–effectiveness analysis. For these predictive distributions and the utility functions introduced in Section 4.3 the optimal treatments are characterized.

Section 4.7 presents a case study with real data that compares four alternative antiretroviral treatments for asymptomatic HIV patients.

Finally, in Sections 4.8 and 4.9 we present the cost–effectiveness acceptability curve for the utility functions defined in Section 4.3. The rationale for this curve is that since the elimination of parameter θ_i from the reward $P_i(c, e | \theta_i)$ depends on the observed samples, this introduces some uncertainty concerning the expectation of the utility function with respect to the predictive distribution $P_i(c, e | \text{data}_i)$, and, consequently, an amount of uncertainty about the optimal treatment. The cost–effectiveness acceptability curve gives us an estimation of this uncertainty.

This elusive curve entails, at the same time, a statistical evaluation of the statistical procedure utilized for the estimation of the predictive reward distribution $P_i(c, e | \text{data}_i)$. In a first reading of this book, these sections can be skipped.

4.2   The net benefit of a treatment

For a generic treatment T with random cost and effectiveness (c, e), the net benefit of (c, e), conditional on a nonnegative parameter R, is a real random variable defined as the linear combination of e and c

$$z(c, e|R) = e\, R - c,$$

where $R \geq 0$ represents the amount of money the decision maker (health provider) is willing to pay for the unit of effectiveness. In what follows, for simplicity in the notation, the dependence of the net benefit z on c and e, conditional on R, will not be explicitly written and we will simply write

$$z = e\, R - c. \tag{4.1}$$

We note that for small values of R the leading term of the net benefit is the second term $(-c)$, and as R grows, the term $e \times R$ becomes the leading term. The net benefit of (c, e) is a slight modification of the incremental net benefit for two treatments introduced by Stinnett and Mullahy (1998).

For every $R \geq 0$ the net benefit transforms the reward space $\mathcal{R} = \{(c, e) : (c, e) \in \mathbb{R}^+ \times \mathcal{E}\}$ into the new reward space $\mathcal{R}(R) = \{z : z \in \mathbb{R}\}$, and, consequently, the space of distributions $\mathcal{P}$ over $\mathcal{R}$ is transformed into a new space of distributions $\mathcal{P}(R)$ over $\mathcal{R}(R)$.

This new space $\mathcal{P}(R)$ is obtained for a given value of the parameter R as follows. The distribution $P(c, e) \in \mathcal{P}$ defines the distribution $P(z|R) \in \mathcal{P}(R)$ by simply doing a change of variables in $P(c, e)$. Indeed, for a fixed R we transform the variables (c, e) into the new variables (z, w) given by

$$\begin{aligned} z &= e\, R - c, \\ w &= e, \end{aligned} \tag{4.2}$$

whose Jacobian is 1. Thus, the distribution $P(c, e)$ can be written as

$P(wR - z, w)$, and the marginal distribution of z is then obtained by integrating out w as

$$P(z|R) = \int_{\mathcal{E}} P(wR - z, w) \, dw. \tag{4.3}$$

The range of z is defined by the transformation (4.2). For instance, for $(c, e) \in \mathbb{R}^+ \times \mathbb{R}$ we have that $z \in \mathbb{R}$.

If the effectiveness space $\mathcal{E}$ is the discrete space $\{e_1, \ldots, e_m\}$, $P(z|R)$ becomes the sum

$$P(z|R) = \sum_{w \in \mathcal{E}} P(wR - z, w), \tag{4.4}$$

and the range of z for a given R is the interval $(-\infty, \max_{i=1,\ldots,m} e_i \times R)$.

4.3   Utility functions of the net benefit

Two utility functions of the net benefit z, conditional on R, will be introduced in this section. In Section 4.3.1 we introduce the utility function $U_1(z|R)$, a linear function of the net benefit z for every R. We remark that $U_1(z|R)$ provides a justification from the decision theory to the well–known *incremental net benefit* (INB) procedure for comparing two alternative treatments (Stinnett and Mullahy, 1998).

In Section 4.3.1.1 we give an interpretation of $U_1(z|R)$ and two criticisms of its use in cost–effectiveness analysis. The main criticism refers the fact that $U_1(z|R)$ takes into account the global net benefit of the patient population, thus detracting from individual net benefit.

In Section 4.3.2 we introduce a second utility function $U_2(z|R)$ which is a nonlinear function of the net benefit z that was originally introduced to avoid the mentioned criticism of the use of $U_1(z|R)$ (Moreno et al., 2010). The meaning of $U_2(z|R)$ is given in Section 4.3.2.1.

4.3.1   The utility function U_1: Optimal treatments

The most used utility function of the net benefit z, conditional on R, is the linear function of z

$$U_1(z|R) = a\,z + b,$$

where $a > 0$ and $b \in \mathbb{R}$ are scale and location parameters. Since the utility function is defined up to a scale and location parameter, in what follows, without loss of generality, we will take $a = 1$ and $b = 0$, and hence

$$U_1(z|R) = z. \tag{4.5}$$

Therefore, the space of probability rewards, conditional on R, is given by the set of distributions

$$\mathcal{P}(R) = \left\{ P_i(z|R) : \mathbb{E}_{P_i}(z|R) = \int_{\mathbb{R}} z\, P_i(z|R)\, dz < \infty,\ i = 1, \dots, k \right\},$$

where the expectation $\mathbb{E}_{P_i}(z|R)$ is the utility of $P_i(z|R)$. This utility is a linear function of the expectation of e and c, that is,

$$\mathbb{E}_{P_i}(z|R) = \mathbb{E}_{P_i}(e)\, R - \mathbb{E}_{P_i}(c).$$

Looking at $\mathbb{E}_{P_i}(z|R)$ as a function of R we see that it is a straight line in the Cartesian plane $(R, \mathbb{E}_{P_i}(z|R))$, whose slope is the expectation of the effectiveness $\mathbb{E}_{P_i}(e)$.

Then, the optimal decision in $\mathcal{D}$, conditional on R, is treatment T_j if the equality

$$\mathbb{E}_{P_j}(z|R) = \max_{i=1,\dots,k} \mathbb{E}_{P_i}(z|R)$$

holds.

Since the optimal treatment is found conditionally on the quantity R, it is convenient to present the set of points R for which decision d_j is optimal under $U_1(z|R)$. This set is given by

$$\mathfrak{R}_j^{U_1} = \left\{ R : \mathbb{E}_{P_j}(z|R) = \max_{i=1,\dots,k} \mathbb{E}_{P_i}(z|R) \right\}, \tag{4.6}$$

for $j = 1, \ldots, k$. Thus, we have that

$$\mathbb{R}^+ = \bigcup_{j=1}^{k} \mathfrak{R}_j^{U_1},$$

where some sets in the class $\left\{ \mathfrak{R}_j^{U_1}, \ j = 1, \ldots, k \right\}$ might be empty.

This way, for any amount of money R the health provider is willing to pay for the unit of effectiveness, the sets $\mathfrak{R}_1^{U_1}, \ldots, \mathfrak{R}_k^{U_1}$ identify the optimal treatment. We can plot the straight lines $\left\{ \mathbb{E}_{P_i}(z|R), \ i = 1, \ldots, k \right\}$ as a function of R to delimit the subsets $\mathfrak{R}_j^{U_1}, \ j = 1, \ldots, k$.

For the case of two treatments T_1 and T_2 with cost and effectiveness expectation $\mathbb{E}_{P_i}(c|R)$ and $\mathbb{E}_{P_i}(e|R)$ for $i = 1, 2$, T_1 is optimal if

$$\mathbb{E}_{P_1}(e)\, R - \mathbb{E}_{P_1}(c) \geq \mathbb{E}_{P_2}(e)\, R - \mathbb{E}_{P_2}(c).$$

Then, assuming that $\mathbb{E}_{P_1}(e) \neq \mathbb{E}_{P_2}(e)$, the set of points R for which T_1 is optimal is given by

$$\mathfrak{R}_1^{U_1} = \begin{cases} \{R : R \leq R^*\}, & \text{if } \mathbb{E}_{P_1}(e) < \mathbb{E}_{P_2}(e), \\[2ex] \{R : R \geq R^*\}, & \text{if } \mathbb{E}_{P_1}(e) > \mathbb{E}_{P_2}(e), \end{cases} \tag{4.7}$$

where

$$R^* = \max\left(0, \ \frac{\mathbb{E}_{P_1}(c) - \mathbb{E}_{P_2}(c)}{\mathbb{E}_{P_1}(e) - \mathbb{E}_{P_2}(e)} \right),$$

and T_2 is optimal for any R in $\mathbb{R}^+ - \mathfrak{R}_1^{U_1}$. If $\mathbb{E}_{P_1}(e) = \mathbb{E}_{P_2}(e)$ then T_1 is optimal if $\mathbb{E}_{P_1}(c) \leq \mathbb{E}_{P_2}(c)$ and T_2 otherwise.

4.3.1.1 Interpretation of the expected utility

An interpretation of the expectation of $U_1(z|R)$ with respect to $P_i(z|R)$ follows from the following approximation. For a fixed value R let $(z_{i1}, \ldots, z_{in})$ be a sample of size n of the net benefit of patients under treatment T_i. This is simply a sample of size n from the net benefit distribution $P_i(z|R)$. From the Law of Large Numbers it follows that for large n we have that

$$\mathbb{E}_{P_i}(z|R) \approx \frac{1}{n}\sum_{j=1}^{n} z_{ij} = R\,\frac{1}{n}\sum_{j=1}^{n} e_{ij} - \frac{1}{n}\sum_{j=1}^{n} c_{ij}.$$

Thus, for a fixed R, the optimal decision under the utility function $U_1(z|R)$ depends on the accumulated patient effectiveness $\sum_{j=1}^{n} e_{ij}$ and cost $\sum_{j=1}^{n} c_{ij}$ of treatment T_i for $i = 1, \ldots, k$, and consequently the effectiveness and cost of the patients under a treatment compensate in total.

While compensation of individual cost is acceptable, compensation of individual effectiveness might not be (Horwitz et al., 1996; Kravitz et al., 2004; Moreno et al., 2010). This first criticism of the use of $U_1(z)$ suggests considering a utility function that does not involve transference of health among patients. A utility function satisfying this requirement is considered in Section 4.3.2.

A second criticism of $U_1(z|R)$ is that the set of rewards $\mathcal{P}(R)$ might be empty. We will go back to this topic in Section 4.6.2.

4.3.2   The utility function U_2: Optimal treatments

For simplicity in presentation, let us first consider the case of comparing two treatments T_1 and T_2 with rewards $P_1(z_1|R)$ and $P_2(z_2|R)$. We assume that the equality $P_1(z_1|z_2, R) = P_1(z_1|R)$ holds, that is, the net benefit of treatments T_1 and T_2 are independent, conditional on R.

For clarity, the random net benefit will be written in capital letters, that is, Z_1 and Z_2, and their values in small letters z_1 and z_2. Let us first introduce the conditional utility function $U_2(z_1|z_2, R)$ for $z_1 \in \mathbb{R}$, whose meaning is the utility function of the net benefit of treatment T_1, conditional on a given net benefit z_2 of treatment T_2 and R. This is defined as

$$U_2(z_1|z_2, R) = \begin{cases} 1, & \text{if } z_1 \geq z_2, \\ 0, & \text{if } z_1 < z_2. \end{cases} \tag{4.8}$$

Similarly, we define $U_2(z_2|z_1, R)$ as

$$U_2(z_2|z_1, R) = \begin{cases} 1, & \text{if } z_2 \geq z_1, \\ 0, & \text{if } z_2 < z_1. \end{cases} \tag{4.9}$$

If we integrate out z_2 in $U_2(z_1|z_2, R)$ with respect to $P_2(z_2|R)$, we obtain the utility of the net benefit of treatment T_1, conditional on R, as

$$U_2(z_1|R) = \int_{z_2 \in \mathbb{R}} 1_{(z_1 \geq z_2)}(z_1, z_2) \, dP_2(z_2|R) = \Pr(Z_2 \leq z_1|R).$$

Then, the expectation of $U_2(z_1|R)$ with respect to $P_1(z_1|R)$, the utility of the reward $P_1(z_1|R)$, turns out to be

$$\mathbb{E}_{P_1} U_2(z_1|R) = \int_{z_1 \in \mathbb{R}} \Pr(Z_2 \leq z_1|R) \, dP_1(z_1|R)$$

$$= \Pr(Z_1 \geq Z_2|R). \tag{4.10}$$

That is, the utility of $P_1(z_1|R)$ is the probability that the net benefit of treatment T_1 is greater than or equal to that of treatment T_2.

Similarly, the expectation of $U_2(z_2|R)$ with respect to $P_2(z_2|R)$, the utility of the reward $P_2(z_2|R)$, is given by

$$\mathbb{E}_{P_2} U_2(z_2|R) = \int_{z_2 \in \mathbb{R}} \int_{z_1 \in \mathbb{R}} 1_{(z_2 \geq z_1)}(z_2, z_1) \, dP_1(z_1|R) \, dP_2(z_2|R)$$

$$= \Pr(Z_2 \geq Z_1|R). \tag{4.11}$$

From (4.10) and (4.11) it follows that, conditional on R, treatment T_1 is preferred to treatment T_2 if the inequality

$$\Pr(Z_1 \geq Z_2|R) \geq \Pr(Z_2 \geq Z_1|R) \tag{4.12}$$

holds.

Therefore, the set of points R for which treatment T_1 is optimal is given by

$$\mathfrak{R}_1^{U_2} = \Big\{ R : \Pr(Z_1 \geq Z_2|R) \geq \Pr(Z_2 \geq Z_1|R) \Big\}, \tag{4.13}$$

and the set of points R for which treatment T_2 is optimal by

$$\mathfrak{R}_2^{U_2} = \left\{ R : \Pr(Z_2 \geq Z_1|R) \geq \Pr(Z_1 \geq Z_2|R) \right\}. \qquad (4.14)$$

It is clear that

$$\mathbb{R}^+ = \mathfrak{R}_1^{U_2} \cup \mathfrak{R}_2^{U_2},$$

and one of the sets $\mathfrak{R}_1^{U_2}$ and $\mathfrak{R}_2^{U_2}$ might be empty.

When the net benefit of treatment T_1 and T_2 are continuous random variables, a simpler expression of the sets $\mathfrak{R}_1^{U_2}$ and $\mathfrak{R}_2^{U_2}$ in (4.13) and (4.14) are given in the next lemma.

Lemma 4.1. *When Z_1 and Z_2 are continuous random variables, the set $\mathfrak{R}_1^{U_2}$ can be written as*

$$\mathfrak{R}_1^{U_2} = \left\{ R : \Pr(Z_1 \geq Z_2|R) \geq 1/2 \right\}, \qquad (4.15)$$

and

$$\mathfrak{R}_2^{U_2} = \mathbb{R}^+ - \mathfrak{R}_1^{U_2}.$$

Proof. From

$$\Pr(Z_1 \geq Z_2|R) + \Pr(Z_2 \geq Z_1|R) = 1$$

we have that the inequality

$$\Pr(Z_1 \geq Z_2|R) \geq \Pr(Z_2 \geq Z_1|R)$$

holds if and only if

$$\Pr(Z_1 \geq Z_2|R) \geq \frac{1}{2}.$$

This proves the assertion.   □

Extension to more than two treatments

The extension of $U_2(z_i|R)$ to the case of $i = 1, \ldots, k$ treatments with $k \geq 3$ is straightforward. For a given R, let $Z_1, \ldots, Z_k$ be the random net benefit of k treatments with distributions $P_1(z_1|R), \ldots, P_k(z_k|R)$, and let Z_{-i} be the random variable defined as

$$Z_{-i} = \max\{Z_1, \ldots, Z_{i-1}, Z_{i+1}, \ldots, Z_k\}$$

with distribution $P(z_{-i}|R)$. We note that this distribution is defined by the distributions $\{P_i(z_i|R), \, i = 1, \ldots, i-1, i+1, \ldots, k\}$. Then, we define the utility U_2 of z_i conditional on z_{-i} and R as

$$U_2(z_i|z_{-i}, R) = \begin{cases} 1, & \text{if } z_i \geq z_{-i}, \\ \\ 0, & \text{if } z_i < z_{-i}, \end{cases}$$

for $i = 1, \ldots, k$. Then, the utility of z_i unconditional on z_{-i} is given by

$$U_2(z_i|R) = \int_{z_i \in \mathbb{R}} 1_{(z_i \geq z_{-i})}(z_{-i}) \, dP(z_{-i}|R) = \Pr(Z_{-i} \leq z_i|R).$$

Therefore, the utility of $P_i(z_i|R)$ is

$$\mathbb{E}_{P_i} U_2(z_i|R) = \int_{z_i \in \mathbb{R}} \Pr(Z_{-i} \leq z_i|R) \, dP_i(z_i|R) = \Pr(Z_i \geq Z_{-i}|R).$$

Consequently, the set of points R for which treatment T_j is optimal is given by

$$\mathfrak{R}_j^{U_2} = \left\{ R : \Pr(Z_j \geq Z_{-j}|R) \geq \max_{i=1,\ldots,k} \Pr(Z_i \geq Z_{-i}|R) \right\}.$$

The utility function $U_2(z_i|R)$ was introduced in Moreno et al. (2010). Applications to real data that use this utility function can be seen in Moreno et al. (2012), where three methadone maintenance programs are compared, Moreno et al. (2013b), which compares two alternative treatments for exacerbated chronic obstrusive pulmonary disease, and Bebu et al. (2016), which compares two treatments for prostate cancer.

4.3.2.1   Interpretation of the expected utility

The interpretation of the expectation of $U_2(z_i|R)$ with respect to $P_i(z_i|R)$ comes from the following argument. For a given R, let $(z_{i1}, \ldots, z_{in_i})$ be a sample of the net benefit of patients under treatment T_i for $i = 1, 2$. Then, for large sample sizes n_1 and n_2 we have that

$$\Pr(Z_1 \geq Z_2|R) \approx \frac{1}{n_1 \cdot n_2} \sum_{k=1}^{n_1} \sum_{j=1}^{n_2} \mathbf{1}_{(z_{1k} \geq z_{2j})}(z_{1k}, z_{2j}).$$

Hence, treatment T_1 is optimal for a given R if for large sample sizes n_1 and n_2 the inequality

$$\frac{1}{n_1 \cdot n_2} \sum_{k=1}^{n_1} \sum_{j=1}^{n_2} \mathbf{1}_{(z_{1k} \geq z_{2j})}(z_{1k}, z_{2j}) \geq \frac{1}{n_1 \cdot n_2} \sum_{k=1}^{n_1} \sum_{j=1}^{n_2} \mathbf{1}_{(z_{2j} \geq z_{1k})}(z_{1k}, z_{2j})$$

holds. That is, for a given R, T_1 is optimal if, for large sample sizes, the proportion of patients for whom the net benefit under T_1 is greater than that of T_2 is larger than the proportion of patients whose net benefit under T_2 is greater than that of T_1. Otherwise, T_2 is optimal.

When Z_1 and Z_2 are continuous random variables the above expression becomes

$$\frac{1}{n_1 \cdot n_2} \sum_{k=1}^{n_1} \sum_{j=1}^{n_2} \mathbf{1}_{(z_{1k} \geq z_{2j})}(z_{1k}, z_{2j}) \geq \frac{1}{2}.$$

We recall that the utility function U_1 does not take into account the number of patients under the treatments but, the total amount of net benefit of patients under the treatments.

We call attention to the fact that the optimal treatment for the utility function U_1 does not necessary coincide with that for the utility function U_2 as the following simple example shows.

Example 4.1. *Let us suppose that the cost of treatments T_1 and T_2 is the same deterministic quantity c_0 and their effectiveness is given by a discrete variable with values $0, 1$ and 2, as a health indicator of bad,* *good and excellent, with distributions*

$$\Pr(e_1) = \begin{cases} 0.1 & if \quad e_1 = 0, \\ 0.5 & if \quad e_1 = 1, \quad and \ \Pr(e_2) = \\ 0.4 & if \quad e_1 = 2, \end{cases} \begin{cases} 0.3 & if \quad e_2 = 0, \\ 0.1 & if \quad e_2 = 1, \\ 0.6 & if \quad e_2 = 2. \end{cases}$$

The reward of treatments T_1 and T_2 are certainly different, although for the utility function $U_1(z|R)$, the utility of $P_1(z_1|R)$ and $P_2(z_2|R)$ is the same, that is,

$$\mathbb{E}_{P_1}(z_1|R) = \mathbb{E}_{P_2}(z_2|R) = 1.3R - c_0.$$

This means that for the utility function $U_1(z|R)$, treatments T_1 and T_2 are equivalent for any R.

However, if we use the utility function $U_2(z|R)$, the utility of $P_1(z_1|R)$ is

$$\Pr(Z_1 \geq Z_2|R) = 0.4 + 0.50 \times 0.4 + 0.1 \times 0.3 = 0.63,$$

and the utility of $P_2(z_2|R)$ is

$$\Pr(Z_2 \geq Z_1|R) = 0.6 + 0.1 \times 0.6 + 0.3 \times 0.1 = 0.69.$$

Therefore, for the utility $U_2(z|R)$, treatment T_2 is preferred to treatment T_1 for any R.

4.4 Penalizing a new treatment

A frequent situation in cost–effectiveness analysis is that of comparing two treatments T_1 and T_2 such that T_1 is a new treatment and T_2 the treatment that is being applied.

Because of the transition cost, it might be reasonable in this case to not choose T_1 unless it is "clearly" preferred to T_2. This implies that for a given utility function $U(z|R)$ of the net benefit, the utility of the reward $P_1(z|R)$ of the new treatment should be penalized by a quantity greater than zero. Consequently, the decision rule is now that treatment T_1 is optimal for a given R, if the inequality

$$\mathbb{E}_{P_1} U(z_1|R) - p \geq \mathbb{E}_{P_2} U(z_2|R)$$

holds, where p is the penalizing term to be fixed by the health provider.

In general, this penalty term can be thought of as a quantification of the additional cost the health system incurs for changing the *status quo* or simply as a consequence of *aversion to the change* of the decision maker.

For the utility function $U_1(z|R)$, the decision rule is that the new treatment T_1 is optimal for any R in the set

$$\mathfrak{R}_1^{U_1}(p_1) = \left\{ R : \mathbb{E}_{P_1}(z_1|R) \geq \mathbb{E}_{P_2}(z_2|R) + p_1 \right\},$$

where the penalizing constant p_1 is such that $0 \leq p_1 < \infty$. Treatment T_2 is optimal for any R in the set $\mathfrak{R}_2^{U_1}(p_1) = \mathbb{R}^+ - \mathfrak{R}_1^{U_1}(p_1)$.

An explicit expression of the optimal treatment in terms of the expectation of the cost and the effectiveness is as follows. For any reward distribution of the cost and the effectiveness $P_i(c, e)$ such that $\mathbb{E}_{P_i}(c) < \infty$ and $\mathbb{E}_{P_i}(e) < \infty$, $i = 1, 2$, and assuming that $\mathbb{E}_{P_1}(e) \neq \mathbb{E}_{P_2}(e)$, treatment T_1 is optimal for any R in the set

$$\mathfrak{R}_1^{U_1}(p_1) = \begin{cases} \{R : R \leq R(p_1)\} & \text{if } \mathbb{E}_{P_1}(e) < \mathbb{E}_{P_2}(e), \\[2mm] \{R : R \geq R(p_1)\} & \text{if } \mathbb{E}_{P_1}(e) > \mathbb{E}_{P_2}(e), \end{cases} \tag{4.16}$$

where

$$R(p_1) = \max\left\{ 0, \frac{\mathbb{E}_{P_1}(c) + p_1 - \mathbb{E}_{P_2}(c)}{\mathbb{E}_{P_1}(e) - \mathbb{E}_{P_2}(e)} \right\}.$$

Treatment T_2 is optimal for any R in the complementary set $\mathfrak{R}_2^{U_1}(p_1) = \mathbb{R}^+ - \mathfrak{R}_1^{U_1}(p_1)$. If $\mathbb{E}_{P_1}(e) = \mathbb{E}_{P_2}(e)$, then T_1 is optimal when $\mathbb{E}_{P_1}(c) + p_1 \leq \mathbb{E}_{P_2}(c)$, and T_2 otherwise.

Notice that when T_1 and T_2 are on an equal footing, $p_1 = 0$ and expression (4.16) coincides with (4.7), thus $\mathfrak{R}_i^{U_1}(0) = \mathfrak{R}_i^{U_1}$, for $i = 1, 2$.

For the utility function $U_2(z|R)$, the set of points R for which T_1 is optimal becomes

$$\mathfrak{R}_1^{U_2}(p_2) = \Big\{ R : \Pr(Z_1 \geq Z_2|R) \geq \Pr(Z_2 \geq Z_1|R) + p_2 \Big\},$$

where the penalizing constant p_2 is such that $0 \leq p_2 \leq 1$. Treatment T_2 is optimal for any R in the set $\mathfrak{R}_2^{U_2}(p_2) = \mathbb{R}^+ - \mathfrak{R}_1^{U_2}(p_2)$.

We note that when Z_1 and Z_2 are continuous random variables, $\mathfrak{R}_1^{U_2}(p_2)$ simplifies to

$$\mathfrak{R}_1^{U_2}(p_2) = \left\{ R : \Pr(Z_1 \geq Z_2|R) \geq \frac{1+p_2}{2} \right\}. \tag{4.17}$$

If $p_2 = 0$ this formula coincides with (4.15), thus $\mathfrak{R}_i^{U_2}(0) = \mathfrak{R}_i^{U_2}$ for $i = 1, 2$.

4.5 Parametric classes of probabilistic rewards

In the preceding sections we assumed that the distribution of the net benefit of treatment T_i was fully determined for $i = 1, \ldots, k$. However, this is not a realistic assumption. The cost–effectiveness analysis typically starts proposing a parametric class of distributions for the cost and the effectiveness of the treatment.

Let $\{P_i(c, e|\theta_i), \theta_i \in \Theta\}$ be the proposed parametric class of distributions of the cost and effectiveness of treatment T_i. This implies that the distribution of the net benefit of treatment T_i,

$$P_i(z|R, \theta_i) = \int_{\mathcal{E}} P_i(e\,R - z, e|\theta_i)\, de,$$

depends not only on R but also on the unknown parameter θ_i. As a consequence, for a given utility function, the optimal treatment is not well defined unless we eliminate the parameter θ_i. Once the parameter is eliminated, the resulting distribution is the one we will use for defining the optimal treatments.

To deal with this situation we assume that there is available a sample of size n_i of the cost and the effectiveness of patients under treatment T_i that we denote as $\mathbf{c}_i = (c_{i1}, \ldots, c_{in_i})$ and $\mathbf{e}_i = (e_{i1}, \ldots, e_{in_i})$. These samples are assumed to be drawn from the distribution $P_i(c, e|\theta_i)$ for the unknown true value θ_i. With the help of these samples we may use either the Bayesian or the frequentist statistical approach for eliminating θ_i from the reward distribution $P_i(c, e|\theta_i)$.

4.5.1 Frequentist predictive distribution of the net benefit

The frequentist approach replaces θ_i in the reward distribution $P_i(c, e|\theta_i)$ with the maximum likelihood estimator $\hat{\theta}_i = \hat{\theta}_i(\mathbf{c}_i, \mathbf{e}_i)$. This is the frequentist way of accommodating the reward of treatment T_i to the samples. The distribution $\widehat{P}_i(c, e|\hat{\theta}_i)$ is the frequentist predictive distribution of (c, e). From this predictive distribution the frequentist predictive distribution of the net benefit is

$$\widehat{P}_i(z|R, \hat{\theta}_i) = \int_{\mathcal{E}} \widehat{P}_i(eR - z, e|\hat{\theta}_i) \, de. \qquad (4.18)$$

We note that the frequentist predictive distribution of the net benefit depends on the data through the MLE estimator but it does not explicitly depend on the sample size n_i.

4.5.2 Bayesian predictive distribution of the net benefit

The Bayesian model is given by the pair $\{P_i(c, e|\theta_i), \pi_i(\theta_i)\}$, where $\pi_i(\theta_i)$ is the prior distribution for θ_i.

In the Bayesian approach, θ_i is eliminated from $P_i(c, e|\theta_i)$ integrating it with respect to the posterior distribution of θ_i, conditional on the samples $\mathbf{c}_i$ and $\mathbf{e}_i$.

Using the random samples $\mathbf{c}_i = (c_{i1}, \ldots, c_{in_i})$ and $\mathbf{e}_i = (e_{i1}, \ldots, e_{in_i})$ coming from the unknown true distribution in the class $\{P_i(c, e|\theta_i), \theta_i \in \Theta\}$, we first update the prior $\pi_i(\theta_i)$ to the posterior distribution $\pi_i(\theta_i|\mathbf{c}_i, \mathbf{e}_i)$. We recall that this posterior distribution of θ_i

is given by

$$\pi_i(\theta_i | \mathbf{c}_i, \mathbf{e}_i) = \frac{\ell_{\mathbf{c}_i, \mathbf{e}_i}(\theta_i) \pi_i(\theta_i)}{m(\mathbf{c}_i, \mathbf{e}_i)},$$

where $\ell_{\mathbf{c}_i, \mathbf{e}_i}(\theta_i) = \prod_{j=1}^{n_i} P_i(c_{ij}, e_{ij} | \theta_i)$ is the likelihood function of parameter θ_i for the random samples $\mathbf{c}_i$ and $\mathbf{e}_i$, and

$$m(\mathbf{c}_i, \mathbf{e}_i) = \int_\Theta \ell_{\mathbf{c}_i, \mathbf{e}_i}(\theta_i) \pi_i(\theta_i) \, d\theta_i$$

is the marginal distribution of the samples $(\mathbf{c}_i, \mathbf{e}_i)$.

Integrating out θ_i from $P_i(c, e | \theta_i)$ with respect to the posterior distribution $\pi_i(\theta_i | \mathbf{c}_i, \mathbf{e}_i)$ we obtain

$$P_i(c, e | \mathbf{c}_i, \mathbf{e}_i) = \int_\Theta P_i(c, e | \theta_i) \pi_i(\theta_i | \mathbf{c}_i, \mathbf{e}_i) \, d\theta_i. \qquad (4.19)$$

This Bayesian predictive distribution accommodates the reward of treatment T_i in the specific samples.

From the predictive distribution $P_i(c, e | \mathbf{c}_i, \mathbf{e}_i)$ the predictive net benefit distribution $P_i(z | R, \mathbf{c}_i, \mathbf{e}_i)$ is obtained as

$$P_i(z | R, \mathbf{c}_i, \mathbf{e}_i) = \int_{\mathcal{E}} P_i(eR - z, e | \mathbf{c}_i, \mathbf{e}_i) \, de. \qquad (4.20)$$

We remark that this distribution depends on the data $(\mathbf{c}_i, \mathbf{e}_i)$ and the prior $\pi_i(\theta_i)$. When $\mathcal{E}$ is discrete, the integral in (4.20) becomes the sum

$$P_i(z | R, \mathbf{c}_i, \mathbf{e}_i) = \sum_{e \in \mathcal{E}} P_i(eR - z, e | \mathbf{c}_i, \mathbf{e}_i).$$

4.6 Statistical models for cost and effectiveness

In this section, for the most frequently used parametric statistical models for the cost and the effectiveness of the treatments, we compute the Bayesian and frequentist predictive distributions of the cost c, the effectiveness e, and the net benefit z. Some models assume that c and e of the treatment are independent, and for some others, a sort of dependency between c and e are modeled. The expression of the sampling models and their mean and variance were given in Chapter 2. The priors of the parameters for defining the Bayesian models will be the objective priors for the sampling models.

For the predictive distributions of the net benefit z we will find the frequentist and Bayesian optimal treatments, conditional on R, for both the linear $U_1(z|R)$ and the nonlinear $U_2(z|R)$ utility functions. For simplicity in illustrating the determination of the optimal treatments, we consider two alternative treatments T_1 and T_2, and it will be assumed that T_1 is the new treatment and T_2 the *status quo*. For the situation where T_1 and T_2 are on an equal footing, the penalizing constants p_1 when using the utility function $U_1(z|R)$ and p_2 when using $U_2(z|R)$ will be set as $p_1 = p_2 = 0$.

4.6.1   The normal–normal model

Let c and e be independent random cost and effectiveness of treatment T_i with normal–normal distribution $\mathcal{N}(c|\mu_i, \sigma_i^2)$ and $\mathcal{N}(e|\eta_i, \tau_i^2)$. For this model and the improper objective reference priors

$$\pi(\mu_i, \sigma_i) = \frac{k_i}{\sigma_i} \mathbf{1}_{\mathbb{R} \times \mathbb{R}^+}(\mu_i, \sigma_i), \quad \pi(\eta_i, \tau_i) = \frac{k_i'}{\tau_i} \mathbf{1}_{\mathbb{R} \times \mathbb{R}^+}(\eta_i, \tau_i),$$

where k_i and k_i' are arbitrary positive constants, the Bayesian model is $\left\{ \mathcal{N}(c|\mu_i, \sigma_i^2) \mathcal{N}(e|\eta_i, \tau_i^2), \pi(\mu_i, \sigma_i) \pi(\eta_i, \tau_i) \right\}$.

Frequentist predictive distribution

For the samples $(\mathbf{c}_i, \mathbf{e}_i)$ the frequentist predictive distribution of c and e is given by

$$\widehat{P}_i(c, e|\mathbf{c}_i, \mathbf{e}_i) = \mathcal{N}(c|\bar{c}_i, s_{c_i}^2) \times \mathcal{N}(e|\bar{e}_i, s_{e_i}^2)$$

where

$$\bar{c}_i = \frac{1}{n_i} \sum_{j=1}^{n_i} c_{ij}, \quad s_{c_i}^2 = \frac{1}{n_i} \sum_{j=1}^{n_i} (c_{ij} - \bar{c}_i)^2,$$

$$\bar{e}_i = \frac{1}{n_i} \sum_{j=1}^{n_i} e_{ij}, \quad s_{e_i}^2 = \frac{1}{n_i} \sum_{j=1}^{n_i} (e_{ij} - \bar{e}_i)^2,$$

are the MLE of $\mu_i, \sigma_i^2, \eta_i$ and τ_i^2.

Further, applying expression (4.3), the frequentist net benefit distribution of treatment T_i turns out to be

$$
\begin{aligned}
\widehat{P}_i(z|R, \mathbf{c}_i, \mathbf{e}_i) &= \int_{-\infty}^{\infty} \mathcal{N}(eR - z|\bar{c}_i, s_{c_i}^2) \mathcal{N}(e|\bar{e}_i, s_{e_i}^2)\, de \\
&= \mathcal{N}(z|\bar{e}_i R - \bar{c}_i, s_{e_i}^2 R^2 + s_{c_i}^2).
\end{aligned} \tag{4.21}
$$

Bayesian predictive distribution

For sample $\mathbf{c}_i$, the Bayesian posterior distribution of (μ_i, σ_i) is given by

$$
\pi(\mu_i, \sigma_i | \mathbf{c}_i) = d \times \mathcal{N}\left(\mu_i | \bar{c}_i, \sigma_i^2 / n_i\right) \times \frac{1}{\sigma_i^{n_i}} \exp\left\{ -\frac{n_i s_{c_i}^2}{2\sigma_i^2} \right\},
$$

where

$$
d = \frac{(n_i s_{c_i}^2)^{(n_i - 1)/2}}{2^{(n_i - 3)/2} \Gamma((n_i - 1)/2)},
$$

and a similar expression for the posterior distribution $\pi(\eta_i, \tau_i | \mathbf{e}_i)$. Then, the Bayesian predictive distribution of c turns out to be

$$
P_i(c|n_i, \bar{c}_i, s_{c_i}) = \frac{1}{\mathrm{B}\left(\frac{n_i - 1}{2}, \frac{1}{2}\right) s_{c_i} \sqrt{n_i + 1}} \left(1 + \frac{1}{n_i + 1} \left(\frac{c - \bar{c}_i}{s_{c_i}} \right)^2 \right)^{-n_i/2}
$$

$$\tag{4.22}$$

where $\mathrm{B}(\alpha, \beta) = \Gamma(\alpha)\Gamma(\beta)/\Gamma(\alpha + \beta)$ denotes the Beta function, and a similar expression for $P_i(e|n_i, \mathbf{e}_i)$. The distribution in (4.22) is recognized as a generalized Student t distribution with location parameter $\bar{c}_i$, scale parameter $s_{c_i} \sqrt{(n_i + 1)/(n_i - 1)}$, and degrees of freedom $n_i - 1$.

From (4.3), the integral expression of the Bayesian predictive distribution of the net benefit of treatment T_i is given by

$$
P_i(z|R, n_i, \mathbf{c}_i, \mathbf{e}_i) = \int_{-\infty}^{\infty} P_i(eR - z|n_i, \mathbf{c}_i) P_i(e|n_i, \mathbf{e}_i)\, de.
$$

This net benefit predictive distribution does not have a closed form expression and hence for every value of R, numerical integration is required. However, its mean is given by $\mathbb{E}_{P_i}(z|R, n_i, \mathbf{c}_i, \mathbf{e}_i) = \bar{e}_i\, R - \bar{c}_i$.

Notice that while the frequentist predictive distribution depends on the MLE $\bar{c}_i$, s_{c_i}, $\bar{e}_i$, s_{e_i}, the Bayesian posterior predictive distribution of the net benefit depends on the MLE plus the sample size n_i.

Example 4.2. *We consider artificial data and normal–normal rewards. The sample means and variances are taken to be $\bar{c} = 19992$, $s_c = 23.15$, $\bar{e} = 2.15$, $s_e = 0.29$, and $n = 10$.*

From (4.21) the frequentist predictive distribution of the net benefit turns out to be

$$\widehat{P}(z|R, \mathbf{c}, \mathbf{e}) = \mathcal{N}(z|2.15R - 19992, 0.29^2 R^2 + 23.15^2).$$

For the above data, the Bayesian predictive distribution $\Pr(z|R, n, \mathbf{c}, \mathbf{e})$ is obtained by simulation from the product of generalized Student t distributions $\Pr(e|R, n, \mathbf{e}) \Pr(c|R, n, \mathbf{c})$. Figure 4.1 shows the Bayesian and frequentist predictive distributions of the net benefit z for two values of R. From this figure we observe that the Bayesian predictive distribution is more spread out than the frequentist.

Table 4.1 displays the means and standard deviations of both the frequentist and Bayesian predictive distributions for three values of R.

TABLE 4.1

Mean and standard deviation of the Bayesian and frequentist predictive distributions of the net benefit for three values of R in Example 4.2

	Bayesian		Frequentist	
	Mean	St. Dev.	Mean	St. Dev.
$R = 5000$	−9248.03	1870.38	−9261.15	1490.51
$R = 10000$	1589.97	3704.72	1469.62	2980.80
$R = 15000$	12297.10	5537.22	12200.40	4471.09

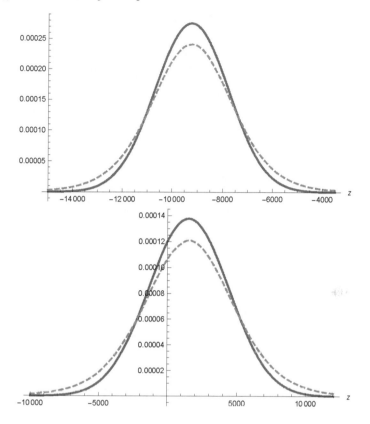

FIGURE 4.1

Bayesian (dashed line) and frequentist (continuous line) predictive distribution of the net benefit in Example 4.2 for $R = 5000$ (upper panel) and $R = 10000$ (lower panel).

Frequentist optimal treatments

The optimal treatment for the utility function $U_1(z|R)$ and penalizing constant p_1 is T_1 for any R in the set

$$\widehat{\mathfrak{R}}_1^{U_1}(p_1) = \left\{ R : (\bar{e}_1 - \bar{e}_2)R - (\bar{c}_1 - \bar{c}_2) \geq p_1 \right\},$$

and T_2 in the set $\widehat{\mathfrak{R}}_2^{U_1}(p_1) = \mathbb{R}^+ - \widehat{\mathfrak{R}}_1^{U_1}(p_1)$. We remark that this is the characterization of the optimal treatments for $U_1(z|R)$ not only for the normal–normal model but for any model such that $(\bar{e}_1 - \bar{e}_2)R - (\bar{c}_1 - \bar{c}_2)$

is the expectation of the frequentist predictive distribution of $Z_1 - Z_2$ conditional on R.

The frequentist predictive distribution of $Z_1 - Z_2$, conditional on R, is the normal distribution with mean $(\bar{e}_1 - \bar{e}_2)R - (\bar{c}_1 - \bar{c}_2)$ and variance $\sum_{i=1}^{2}(s_{e_i}^2 R^2 + s_{c_i}^2)$. Then, for $U_2(z|R)$ and penalizing constant p_2, treatment T_1 is optimal for any R in the set

$$\widehat{\mathfrak{R}}_1^{U_2}(p_2) = \left\{ R : \Phi\left(\frac{(\bar{e}_1 - \bar{e}_2)R - (\bar{c}_1 - \bar{c}_2)}{\sqrt{\sum_{i=1}^{2}(s_{e_i}^2 R^2 + s_{c_i}^2)}}\right) \geq \frac{1 + p_2}{2} \right\},$$

and T_2 for any R in $\widehat{\mathfrak{R}}_2^{U_2}(p_2) = \mathbb{R}^+ - \widehat{\mathfrak{R}}_1^{U_2}(p_2)$. Here, $\Phi(x)$ denotes the cumulative standard normal distribution at point x.

Example 4.3. *Let us consider the artificial data sets of cost and effectiveness displayed in Table 4.2.*

TABLE 4.2

Data in Example 4.3.

Treatment	n_i	Cost		Effectiveness	
		$\bar{c}_i$	s_{c_i}	$\bar{e}_i$	s_{e_i}
T_1	10	5224.64	370.84	0.203	0.09
T_2	10	4297.39	715.59	0.10	0.009

For that data we plot in Figure 4.2

$$\mathbb{E}_{\widehat{P}_1}(z|R, data) - \mathbb{E}_{\widehat{P}_2}(z|R, data) = (\bar{e}_1 - \bar{e}_2)R - (\bar{c}_1 - \bar{c}_2),$$

and

$$\widehat{\Pr}(Z_1 \geq Z_2|R, data) = \Phi\left(\frac{(\bar{e}_1 - \bar{e}_2)R - (\bar{c}_1 - \bar{c}_2)}{\sqrt{\sum_{i=1}^{2}(s_{e_i}^2 R^2 + s_{c_i}^2)}}\right).$$

From those expressions we immediately obtain $\widehat{\mathfrak{R}}_1^{U_1}(p_1)$ and $\widehat{\mathfrak{R}}_1^{U_2}(p_2)$.

From Figure 4.2 it follows that in an equal–footing scenario where $p_1 = p_2 = 0$, and due to the symmetry of the normal–normal model, we have that $\widehat{\mathfrak{R}}_1^{U_1} = \widehat{\mathfrak{R}}_1^{U_2}$. For the above data, $\widehat{\mathfrak{R}}_1^{U_1} = \{R : R \geq 9002\}$.

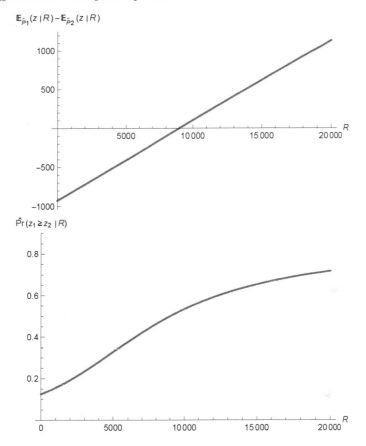

FIGURE 4.2

Upper panel: $\mathbb{E}_{\widehat{P}_1}(z|R, \text{data}) - \mathbb{E}_{\widehat{P}_2}(z|R, \text{data})$ as a function of R in Example 4.3. **Lower panel:** $\widehat{\Pr}(Z_1 \geq Z_2|R, \text{data})$ as a function of R in Example 4.3.

Bayesian optimal treatments

For any penalizing constant p_1 and the utility function $U_1(z|R)$, the set of points R for which T_1 is the Bayesian optimal treatment satisfies $\mathfrak{R}_1^{U_1}(p_1) = \widehat{\mathfrak{R}}_1^{U_1}(p_1)$. This is because the means of the Bayesian and frequentist predictive distributions coincide in the normal–normal model.

However, for the utility function $U_2(z|R)$, the optimal treatment for the frequentist and the Bayesian predictive distributions can differ for some values of R. The following example gives a numerical illustration of that.

Example 4.3 (continued). *The probabilities* $\widehat{\Pr}(Z_1 \geq Z_2 | R, \mathbf{c}_i, \mathbf{e}_i)$ *and* $\Pr(Z_1 \geq Z_2 | R, n_i, \mathbf{c}_i, \mathbf{e}_i)$ *for the data in Table 4.2 are displayed in Figure 4.3 as a function of* R.

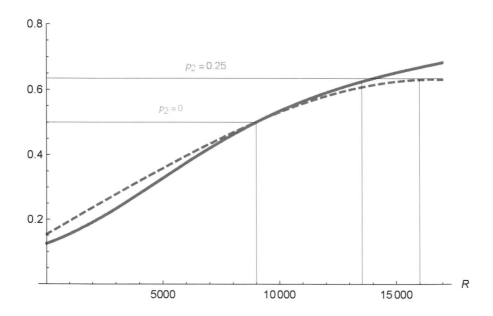

FIGURE 4.3

Frequentist (continuous line) and Bayesian (dashed line) predictive distribution of $Z_1 - Z_2$ as a function of R in Example 4.3.

For $p_2 = 0$ *and* $p_2 = 0.25$ *the frequentist and Bayesian optimal treatments for the utility function* U_2 *are given in Table 4.3*

4.6.2   The lognormal–normal model

Let us assume that c and e are independent random cost and effectiveness of treatment T_i with lognormal–normal distribution $\Lambda(c|\mu_i, \sigma_i^2) \times \mathcal{N}(e|\eta_i, \tau_i^2)$, where $\Lambda(c|\mu_i, \sigma_i^2)$ denotes the lognormal distribution given in (2.17).

TABLE 4.3

Sets of R for which treatment T_1 is optimal for the utility U_2 and $p_2 = 0, p_2 = 0.25$.

Bayesian	Frequentist
$\mathfrak{R}_1^{U_2}(0) = \{R : R \geq 9002\}$	$\widehat{\mathfrak{R}}_1^{U_2}(0) = \{R : R \geq 9002\}$
$\mathfrak{R}_1^{U_2}(0.25) = \{R : R \geq 16036.70\}$	$\widehat{\mathfrak{R}}_1^{U_2}(0.25) = \{R : R \geq 13537.30\}$

For this model the objective reference priors are

$$\pi(\mu_i, \sigma_i) = \frac{k_i}{\sigma_i} \mathbf{1}_{\mathbb{R} \times \mathbb{R}+}(\mu_i, \sigma_i), \quad \pi(\eta_i, \tau_i) = \frac{k_i'}{\tau_i} \mathbf{1}_{\mathbb{R} \times \mathbb{R}+}(\eta_i, \tau_i),$$

where k_i and k_i' are arbitrary positive constants and the Bayesian model is $\left\{ \Lambda(c|\mu_i, \sigma_i^2) \mathcal{N}(e|\eta_i, \tau_i^2), \pi(\mu_i, \sigma_i)\pi(\eta_i, \tau_i) \right\}$.

Frequentist predictive distribution

For the samples $(\mathbf{c}_i, \mathbf{e}_i)$ the joint frequentist predictive distribution of c and e is given by

$$\widehat{P}_i(c, e|\mathbf{c}_i, \mathbf{e}_i) = \Lambda(c|\tilde{c}_i, \tilde{s}_{c_i}^2) \times \mathcal{N}(e|\bar{e}_i, s_{e_i}^2),$$

where

$$\tilde{c}_i = \frac{1}{n_i} \sum_{j=1}^{n_i} \log(c_{ij}), \quad \tilde{s}_{c_i}^2 = \frac{1}{n_i} \sum_{j=1}^{n_i} (\log(c_{ij}) - \tilde{c}_i)^2.$$

Then, the integral expression of the frequentist net benefit distribution is given by

$$\widehat{P}_i(z|R, \mathbf{c}_i, \mathbf{e}_i) = \int_{-\infty}^{\infty} \Lambda(eR - z|\tilde{c}_i, \tilde{s}_{c_i}^2) \mathcal{N}(e|\bar{e}_i, s_{e_i}^2) \, de.$$

In contrast with the normal–normal model, this frequentist predictive distribution doesn't have a closed form expression and needs numerical integration to be evaluated for every R. However, the expectation of z is easily obtained as it is a linear function of c and e, and hence it turns out to be

$$\mathbb{E}_{\widehat{P}_i}(z|R) = \bar{e}_i R - \exp\left\{\tilde{c}_i + \frac{\tilde{s}_{c_i}^2}{2}\right\}.$$

Bayesian predictive distribution

The Bayesian predictive distribution of e for treatment T_i, conditional on the data $\mathbf{e}_i = (e_{i1}, \ldots, e_{in_i})$, is the generalized Student t distribution $P_i(e|n_i, \bar{e}_i, s_{e_i})$ given in (4.22) interchanging c, $\bar{c}_i$, s_{c_i} with e, $\bar{e}_i$, s_{e_i}.

For the lognormal distribution $\Lambda(c|\mu_i, \sigma_i^2)$, and the reference prior $\pi(\mu_i, \sigma_i) \propto 1/\sigma_i \mathbf{1}_{\mathbb{R}\times\mathbb{R}^+}(\mu_i, \sigma_i)$, the Bayesian predictive distribution of c, conditional on the data $\mathbf{c}_i = (c_{i1}, \ldots, c_{in_i})$ turns out to be the logStudent t distribution

$$P_i(c|n_i, \tilde{c}_i, \tilde{s}_{c_i}) = A \frac{1}{c} \left(1 + \frac{1}{n_i + 1}\left(\frac{\log(c) - \tilde{c}_i}{\tilde{s}_{c_i}}\right)^2\right)^{-n_i/2}, \quad (4.23)$$

where

$$A = \frac{1}{\tilde{s}_{c_i}\sqrt{n_i + 1}B((n_i - 1)/2, 1/2)}.$$

Then, the Bayesian predictive distribution of (c, e) is given by

$$P_i(c, e|n_i, \mathbf{c}_i, \mathbf{e}_i) = P_i(c|n_i, \tilde{c}_i, \tilde{s}_{c_i}) \times P_i(e|n_i, \bar{e}_i, s_{e_i}). \quad (4.24)$$

Using expression (4.24) we can simulate the Bayesian predictive distribution of $z = eR - c$, for every value of R.

Example 4.4. *We consider artificial data sets and lognormal–normal rewards for a generic treatment T. The sample means and variances are taken to be $\tilde{c} = 9.08$, $\tilde{s}_c = 1.22$, $\bar{e} = 3.01$, $s_e = 0.08$, and $n = 10$.*

Figure 4.4 shows the frequentist and Bayesian predictive distributions of the net benefit for $R = 5000$ euros, from which it can be seen that the tails of the Bayesian are thicker than that of the frequentist.

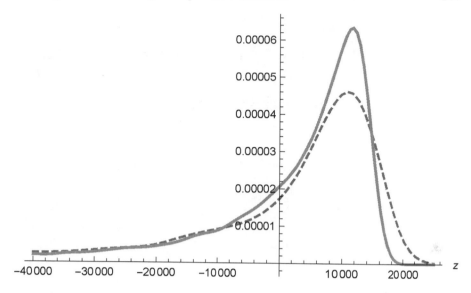

FIGURE 4.4

Bayesian (dashed line) and frequentist (continuous line) predictive distribution of the net benefit in Example 4.4 for $R = 5000$.

Frequentist optimal treatments

For the utility function $U_1(z|R)$ and penalizing constant p_1, treatment T_1 is optimal for any R in the set

$$\widehat{\mathfrak{R}}_1^{U_1}(p_1) = \left\{ R : (\bar{e}_1 - \bar{e}_2)R - d_{12} \geq p_1 \right\},$$

where

$$d_{12} = \exp\left\{ \tilde{c}_1 + \frac{\tilde{s}_{c_1}^2}{2} \right\} - \exp\left\{ \tilde{c}_2 + \frac{\tilde{s}_{c_2}^2}{2} \right\},$$

and T_2 for any R in the set $\widehat{\mathfrak{R}}_2^{U_1}(p_1) = \mathbb{R}^+ - \widehat{\mathfrak{R}}_1^{U_1}(p_1)$.

Further, for the utility function $U_2(z|R)$ and penalizing constant p_2, the optimal treatment is T_1 for any R in the set

$$\widehat{\mathfrak{R}}_1^{U_2}(p_2) = \left\{ R : \widehat{\mathrm{Pr}}(Z_1 \geq Z_2 | R, \mathbf{c}_1, \mathbf{e}_1, \mathbf{c}_2, \mathbf{e}_2) \geq \frac{1 + p_2}{2} \right\},$$

where the probability is computed using the frequentist predictive distributions $\widehat{P}_1(z|R, \mathbf{c}_1, \mathbf{e}_1)$ and $\widehat{P}_2(z|R, \mathbf{c}_2, \mathbf{e}_2)$.

Bayesian optimal treatments

Unfortunately, the logStudent t distribution $P_i(c|n_i, \tilde{c}_i, \tilde{s}_{c_i})$ in (4.23) does not have a mean as shown in Lemma 4.2.

Lemma 4.2. *The logStudent t distribution $P(c|n, \tilde{c}, \tilde{s})$ has no moments.*

Proof. The expectation of the logStudent distribution is such that

$$\mathbb{E}(c|n, \tilde{c}, \tilde{s}) = \int_0^\infty c\, P(c|n, \tilde{c}, \tilde{s})\, dc$$

$$= A \int_0^\infty \left(1 + \frac{(\log(c) - \tilde{c})^2}{(n+1)\tilde{s}^2}\right)^{-n/2} dc$$

$$\geq A \int_0^K \left(1 + \frac{(\log(c) - \tilde{c})^2}{(n+1)\tilde{s}^2}\right)^{-n/2} dc,$$

where K is an arbitrary positive real number. For $c \geq c_0$ the integrand is a decreasing function of c, and hence for large K we have that

$$\int_0^K \left(1 + \frac{(\log(c) - \tilde{c})^2}{(n+1)\tilde{s}^2}\right)^{-n/2} dc \geq K \left(1 + \frac{(\log(K) - \tilde{c})^2}{(n+1)\tilde{s}^2}\right)^{-n/2}.$$

Taking the limit in the last inequality as $K \to \infty$ we have that

$$\lim_{K \to \infty} \int_0^K \left(1 + \frac{(\log(c) - \tilde{c})^2}{(n+1)\tilde{s}^2}\right)^{-n/2} dc \geq \lim_{K \to \infty} K \left(1 + \frac{(\log(K) - \tilde{c})^2}{(n+1)\tilde{s}^2}\right)^{-n/2}$$

$$= \infty.$$

Since $A > 0$, we have that

$$\mathbb{E}(c|n, \tilde{c}, \tilde{s}) = \int_0^\infty c\, P(c|n, \tilde{c}, \tilde{s})\, dc = \infty.$$

This proves the assertion. $\square$

Lemma 4.2 implies that the utility function $U_1(z|R)$ cannot be utilized if the cost is modeled by a lognormal distribution and an objective Bayesian analysis is carried out.

When the computation is made by simulation, and we are not aware of the result in Lemma 4.2, we obtain a number for the expectation of

$U_1(z|R)$ for every R, which gives the false impression of the existence of the mean. The warning is that we first need to check whether the expectation of c and e exist there for the model we are using.

For the utility function $U_2(z|R)$ and penalizing constant p_2, the Bayesian optimal treatment is T_1 for any R in the set

$$\mathfrak{R}_1^{U_2}(p_2) = \left\{ R : \Pr(Z_1 \geq Z_2 | R, \mathbf{c}_1, \mathbf{e}_1, \mathbf{c}_2, \mathbf{e}_2) \geq \frac{1 + p_2}{2} \right\},$$

where the probability is computed using $P_i(c|n_i, \tilde{c}_i, \tilde{s}_{c_i}^2)$ and $P_i(e|n_i, \bar{e}_i, s_{e_i}^2)$ for $i = 1, 2$.

Example 4.5. *We generate two data sets of size $n_1 = n_2 = 10$ from $\Lambda(c|9, 0.25) \times \mathcal{N}(e|3, 0.01)$ and $\Lambda(c|8.50, 4) \times \mathcal{N}(e|3, 0.04)$ that are assumed to be the reward distributions of treatment T_1 and T_2, respectively. Means and standard deviations of these data are presented in Table 4.4.*

TABLE 4.4

Sample means and standard deviations in Example 4.5.

Treatment	n_i	Cost		Effectiveness	
		$\tilde{c}_i$	$\tilde{s}_{c_i}$	$\bar{e}_i$	s_{e_i}
T_1	10	9.20	0.49	3.02	0.10
T_2	10	8.86	1.85	2.78	0.15

The sets of R for which each treatment is optimal for the utility $U_1(z|R)$ and $U_2(z|R)$ are presented in Table 4.5 for $p_1 = 0$ and $p_2 = 0$.

4.6.3 The lognormal–Bernoulli model

Let us assume that c and e are independent random cost and effectiveness of treatment T_i with lognormal–Bernoulli distribution $\Lambda(c|\mu_i, \sigma_i^2) \times \text{Be}(e|\theta_i)$, where $\text{Be}(e|\theta_i)$ denotes the Bernoulli distribution given in (2.11).

The prior for (μ_i, σ_i) is the objective reference prior $\pi(\mu_i, \sigma_i) \propto$

TABLE 4.5

$\mathfrak{R}_1^{U_j}$ and $\widehat{\mathfrak{R}}_1^{U_j}$ sets in Example 4.3 for $j = 1, 2$.

Utility	Bayesian	Frequentist
U_1	It does not exist	$\widehat{\mathfrak{R}}_1^{U_1} = \mathbb{R}^+$
U_2	$\mathfrak{R}_1^{U_2} = \{R : R \geq 8507.15\}$	$\widehat{\mathfrak{R}}_1^{U_2} = \{R : R \geq 9671.41\}$

$1/\sigma_i \mathbf{1}_{\mathbb{R} \times \mathbb{R}^+}(\mu_i, \sigma_i)$ and for θ_i it is assigned the Jeffreys proper prior distribution

$$\pi(\theta_i) = \frac{1}{\pi} \theta_i^{-1/2} (1 - \theta_i)^{-1/2}, \quad 0 < \theta_i < 1.$$

Hence the Bayesian model is $\left\{ \Lambda(c|\mu_i, \sigma_i^2) \, \mathrm{Be}(e|\theta_i), \pi(\mu_i, \sigma_i)\pi(\theta_i) \right\}$.

Frequentist predictive distribution

For the samples $(\mathbf{c}_i, \mathbf{e}_i)$ the frequentist predictive distribution of c and e for treatment T_i is given by

$$\widehat{P}_i(c, e|\mathbf{c}_i, \mathbf{e}_i) = \Lambda(c|\tilde{c}_i, \tilde{s}_{c_i}^2) \times \mathrm{Be}(e|\bar{e}_i).$$

Therefore, from (4.4) it follows that the frequentist net benefit predictive distribution turns out to be

$$\begin{aligned}
\widehat{P}_i(z|R, \mathbf{c}_i, \mathbf{e}_i) &= \sum_{e=0}^{1} \Lambda(eR - z|\tilde{c}_i, \tilde{s}_{c_i}^2) \times \mathrm{Be}(e|\bar{e}_i) \\
&= (1 - \bar{e}_i) \, \Lambda(-z|\tilde{c}_i, \tilde{s}_{c_i}^2) \mathbf{1}_{(-\infty,0)}(z) \\
&\quad + \bar{e}_i \, \Lambda(R - z|\tilde{c}_i, \tilde{s}_{c_i}^2) \mathbf{1}_{(-\infty,R)}(z). \quad (4.25)
\end{aligned}$$

Bayesian predictive distribution

The Bayesian predictive distribution of c, conditional on $n_i, \tilde{c}_i, \tilde{s}_{c_i}^2$, is the logStudent t distribution given in (4.23), and the Bayesian predictive distribution of e, conditional on n_i and $\bar{e}_i$, is given by

$$\Pr(e|n_i, \bar{e}_i) = \frac{n_i(1 - \bar{e}_i) + 0.5}{n_i + 1} \mathbf{1}_{(e=0)}(e) + \frac{n_i \bar{e}_i + 0.5}{n_i + 1} \mathbf{1}_{(e=1)}(e). \quad (4.26)$$

The difference between the Bayesian predictive distribution $\Pr(e|n_i, \bar{e}_i)$ and the frequentist $\mathrm{Be}(e|\bar{e}_i)$ becomes negligible when n_i is large. For instance, for $n_i = 5$ and $\bar{e}_i = 0.2$ we have

$$|\Pr(e = 0|5, 0.2) - \mathrm{Be}(e = 0|0.2)| = 0.05,$$

and for $n_i = 100$ and $\bar{e}_i = 0.2$ the absolute value of the difference is as small as 0.003.

It is easy to show that when sampling from $\mathrm{Be}(e|\theta)$ we have that

$$\lim_{n_i \to \infty} \Pr(e|n_i, \bar{e}_i) = \mathrm{Be}(e|\theta), \qquad [P_\theta].$$

The Bayesian predictive net benefit distribution turns out to be

$$
\begin{aligned}
P_i(z|R, n_i, \mathbf{c}_i, \mathbf{e}_i) &= \sum_{e=0}^{1} P_i(eR - z|n_i, \tilde{c}_i, \tilde{s}_{c_i}^2) \Pr(e|n_i, \bar{e}_i) \\
&= \frac{n_i(1 - \bar{e}_i) + 0.5}{n_i + 1} P_i(-z|n_i, \tilde{c}_i, \tilde{s}_{c_i}^2) \mathbf{1}_{(-\infty, 0)}(z) \\
&\quad + \frac{n_i \bar{e}_i + 0.5}{n_i + 1} P_i(R - z|n_i, \tilde{c}_i, \tilde{s}_{c_i}^2) \mathbf{1}_{(-\infty, R)}(z),
\end{aligned}
$$

$$(4.27)$$

where P_i denotes the logStudent t distribution given in (4.23). The next example illustrates the frequentist and Bayesian predictive distributions for this model.

Example 4.6. *For the lognormal–Bernoulli model and artificial data with $n = 10$, $\tilde{c} = 9.13$, $\tilde{s}_c = 0.09$ and $\bar{e} = 0.7$, Figure 4.5 displays the Bayesian and frequentist predictive net benefit distributions for $R = 5000$. They have two local modes located at points -9141.40 and -4146.13 for the Bayesian, and -9153.54 and -4153.58 for the frequentist.*

From Figure 4.5 it follows again that the tails of the Bayesian predictive distribution are thicker than those of the frequentist.

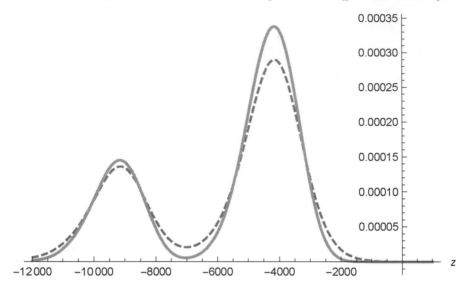

FIGURE 4.5

Bayesian (dashed line) and frequentist (continuous line) predictive distributions of the net benefit for $R = 5000$ in Example 4.6.

Frequentist optimal treatments

For the utility function $U_1(z|R)$ and penalizing constant p_1, treatment T_1 is optimal for any R in the set

$$\widehat{\mathfrak{R}}_1^{U_1}(p_1) = \left\{ R : (\bar{e}_1 - \bar{e}_2)R - d_{12} \geq p_1 \right\},$$

where

$$d_{12} = \exp\left\{ \tilde{c}_1 + \frac{\tilde{s}_{c_1}^2}{2} \right\} - \exp\left\{ \tilde{c}_2 + \frac{\tilde{s}_{c_2}^2}{2} \right\},$$

and T_2 for any R in the set $\widehat{\mathfrak{R}}_2^{U_1}(p_1) = \mathbb{R}^+ - \widehat{\mathfrak{R}}_1^{U_1}(p_1)$.

Further, for the utility function $U_2(z|R)$ and penalizing constant p_2, the optimal treatment is T_1 for any R in the set

$$\widehat{\mathfrak{R}}_1^{U_2}(p_2) = \left\{ R : \widehat{\mathrm{Pr}}(Z_1 \geq Z_2 | R, \mathbf{c}_1, \mathbf{e}_1, \mathbf{c}_2, \mathbf{e}_2) \geq \frac{1 + p_2}{2} \right\},$$

where the probability is computed using the frequentist predictive distributions $\widehat{P}_1(z|R, \mathbf{c}_1, \mathbf{e}_1)$ and $\widehat{P}_2(z|R, \mathbf{c}_2, \mathbf{e}_2)$ given in (4.25) for $i = 1, 2$.

Bayesian optimal treatments

For the utility function $U_1(z|R)$, Lemma 4.2 shows that the Bayesian optimal treatment does not exist.

For the utility function $U_2(z|R)$ and penalizing constant p_2, the Bayesian optimal treatment is T_1 for any R in the set

$$\mathfrak{R}_1^{U_2}(p_2) = \left\{ R : \Pr(Z_1 \geq Z_2 | R, n_1, n_2, \mathbf{c}_1, \mathbf{e}_1, \mathbf{c}_2, \mathbf{e}_2) \geq \frac{1 + p_2}{2} \right\},$$

where the probability is computed using $P_i(z|R, n_i, \mathbf{c}_i, \mathbf{e}_i)$ for $i = 1, 2$, in (4.27).

Example 4.7. *We simulate two data sets with size $n_1 = n_2 = 20$ of c and e from two lognormal–Bernoulli distributions with parameters $\mu_1 = 3$, $\sigma_1^2 = 2$, $\theta_1 = 0.7$ and $\mu_2 = 6$, $\sigma_2^2 = 4$, $\theta_2 = 0.9$. These distributions are the rewards of treatment T_1 and T_2, respectively.*

Sample means and standard deviations of the simulated data are presented in Table 4.6.

TABLE 4.6

Sample means and standard deviations in Example 4.7.

Treatment	n_i	Cost		Effectiveness	
		$\tilde{c}_i$	$\tilde{s}_{c_i}$	$\bar{e}_i$	s_{e_i}
T_1	20	2.71	1.34	0.55	0.50
T_2	20	5.94	1.98	0.90	0.30

For these data we compute the set of points R for which treatment T_1 is optimal for the utility functions $U_1(z|R)$ and $U_2(z|R)$.

In an equal–footing scenario where $p_1 = p_2 = 0$, Figure 4.6 displays $\mathbb{E}_{\widehat{P}_i}(z_i | R, data_i)$ for $i = 1, 2$, and $\widehat{\Pr}(Z_1 \geq Z_2 | R, data)$ as a function of R. From the upper panel in Figure 4.6 it follows that

$$\widehat{\mathfrak{R}}_1^{U_1} = \{ R : R \leq 7646.65 \}.$$

From the lower panel we conclude that $\widehat{\mathfrak{R}}_1^{U_2} = \mathbb{R}^+$.

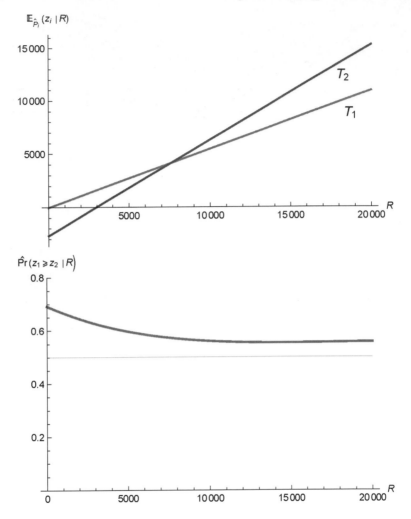

FIGURE 4.6

Upper panel: $\mathbb{E}_{\widehat{P}_i}(z_i|R,\text{data}_i)$, $i=1,2$, as a function of R for the lognormal–Bernoulli model. **Lower panel:** $\widehat{\Pr}(Z_1 \geq Z_2|R,\text{data})$ as a function of R for the lognormal–Bernoulli model.

Table 4.7 displays the sets of R for which treatment T_1 is optimal for the utility functions $U_1(z|R)$ and $U_2(z|R)$. The result using the Bayesian predictive distribution and $U_2(z|R)$ is based on values generated from the net benefit distribution in (4.27).

TABLE 4.7

Sets of R for which T_1 is optimal under the Bayesian and frequentist approaches in Example 4.7.

Utility	Bayesian	Frequentist
U_1	It does not exist	$\widehat{\mathfrak{R}}_1^{U_1} = \{R : R \leq 7646.65\}$
U_2	$\mathfrak{R}_1^{U_2} = \mathbb{R}^+$	$\widehat{\mathfrak{R}}_1^{U_2} = \mathbb{R}^+$

4.6.4 The bivariate normal model

Let (c, e) be the random cost and effectiveness of treatment T_i that follows the bivariate normal distribution $\mathcal{N}_2((c, e)^\top | (\mu_i, \eta_i)^\top, \Sigma_i)$, where $(c, e)^\top$ denotes the transpose of (c, e), $(\mu_i, \eta_i)^\top$ is the mean vector and

$$\Sigma_i = \begin{pmatrix} \sigma_i^2 & \phi_i \\ \phi_i & \tau_i^2 \end{pmatrix}$$

is the covariance matrix, with σ_i^2 and τ_i^2 the variances of the cost and effectiveness, and ϕ_i the covariance of cost and effectiveness. Let $(\mathbf{c}_i, \mathbf{e}_i)$ be a sample of size n_i from this bivariate normal distribution.

The following lemma is useful in obtaining the distribution of the net benefit.

Lemma 4.3. *Let* $\mathbf{X}$ *be a random vector of dimension p following the normal distribution with mean* $\boldsymbol{\mu} = (\mu_1, \ldots, \mu_p)^\top$ *and covariance matrix* $\mathbf{V}$ *of dimensions $p \times p$, invertible and positive definite. Then, the random vector* $\mathbf{Y} = \mathbf{a} + \mathbf{B}\mathbf{X}$*, where $\mathbf{a}$ is a vector of dimension q, $q \leq p$, and $\mathbf{B}$ is a matrix of dimensions $q \times p$, follows the $q-$variate normal distribution with mean* $\mathbf{a} + \mathbf{B}\boldsymbol{\mu}$ *and covariance matrix* $\mathbf{B}\mathbf{V}\mathbf{B}^\top$.

For a given $R \geq 0$, the distribution of z is obtained by applying Lemma 4.3 for $q = 1$, $p = 2$, $\mathbf{X} = (c, e)^\top$, $\mathbf{a} = \mathbf{0}$, and $\mathbf{B} = (-1, R)$. Hence, the distribution of the net benefit z is the normal distribution $\mathcal{N}(z | \eta R - \mu, \tau^2 R^2 + \sigma^2 - 2R\phi)$.

Frequentist predictive distribution

The frequentist net benefit predictive distribution is the normal

$$\widehat{P}_i(z|R, \mathbf{c}_i, \mathbf{e}_i) = \mathcal{N}(z|\bar{e}_i R - \bar{c}_i, s_{e_i}^2 R^2 + s_{c_i}^2 - 2Rs_{c_i e_i}), \qquad (4.28)$$

where $s_{c_i}^2$ and $s_{e_i}^2$ are the sample variances of $\mathbf{c}_i$ and $\mathbf{e}_i$, and

$$s_{c_i e_i} = \frac{1}{n_i} \sum_{j=1}^{n_i} (c_{ij} - \bar{c}_i)(e_{ij} - \bar{e}_i).$$

Bayesian predictive distribution

For the improper Jeffreys prior for the parameters, that is,

$$\pi((\mu_i, \eta_i), \Sigma_i) \propto |\Sigma_i|^{-3/2},$$

some algebra shows (Gelman et al., 2004) that the posterior distribution of (μ_i, η_i) is the bivariate Student t distribution

$$\pi(\mu_i, \eta_i | \mathbf{c}_i, \mathbf{e}_i) = \mathcal{T}_2\left((\mu_i, \eta_i)^\top | (\bar{c}_i, \bar{e}_i)^\top, \frac{1}{n_i - 2} S_i, n_i - 2\right),$$

and the posterior distribution of the matrix Σ_i is the multivariate inverted Wishart distribution

$$\pi(\Sigma_i | \mathbf{c}_i, \mathbf{e}_i) = \mathcal{IW}\left(\Sigma_i | n_i S_i, n_i - 1\right),$$

where S_i, the MLE of the covariance matrix Σ_i, is given by the matrix

$$S_i = \begin{pmatrix} s_{c_i}^2 & s_{c_i e_i} \\ s_{c_i e_i} & s_{e_i}^2 \end{pmatrix}.$$

Thus, the Bayesian predictive distribution of c and e becomes the bivariate Student t distribution

$$P_i(c, e | \mathbf{c}_i, \mathbf{e}_i) = \mathcal{T}_2\left((c, e)^\top | (\bar{c}_i, \bar{e}_i)^\top, \frac{n_i + 1}{n_i - 2} S_i, n_i - 2\right). \qquad (4.29)$$

The Bayesian predictive net benefit distribution $P_i(z|R, \mathbf{c}_i, \mathbf{e}_i)$ does not have a close form expression, although a sample from it can be easily obtained by simulation from $P_i(c, e | \mathbf{c}_i, \mathbf{e}_i)$.

Example 4.2 (continued). *We consider the artificial data sets in Example 4.2 plus $s_{ce} = 3.36$. The frequentist predictive distribution of the net benefit of treatment T, conditional on R, turns out to be*

$$\widehat{P}(z|R, \mathbf{c}, \mathbf{e}) = \mathcal{N}(z|2.15R - 19992, 0.29^2 R^2 + 23.15^2 - 6.72R)$$

and the Bayesian predictive distribution of (c, e)

$$P(c, e|\mathbf{c}, \mathbf{e}) = \mathcal{T}_2((c, e)^\top|(19992, 2.15)^\top, 1.37S, 8),$$

where

$$S = \begin{pmatrix} 23.15^2 & 3.36 \\ 3.36 & 0.29^2 \end{pmatrix}.$$

Frequentist optimal treatments

The optimal treatment in $\{T_1, T_2\}$ for the utility function $U_1(z|R)$ only depends on the sample means $\bar{c}_i$ and $\bar{e}_i$, and hence the analysis coincides with that for the normal–normal model in Section 4.6.1.

For the utility function $U_2(z|R)$, and penalizing constant p_2, treatment T_1 is optimal for any R in the set

$$\widehat{\mathfrak{R}}_1^{U_2}(p_2) = \left\{ R : \Phi\left(\frac{(\bar{e}_1 - \bar{e}_2)R - (\bar{c}_1 - \bar{c}_2)}{\sqrt{\sum_{i=1}^2 (s_{e_i}^2 R^2 + s_{c_i}^2 - 2Rs_{c_i e_i})}} \right) \geq \frac{1 + p_2}{2} \right\},$$

and T_2 for any R in $\widehat{\mathfrak{R}}_2^{U_2}(p_2) = \mathbb{R}^+ - \widehat{\mathfrak{R}}_1^{U_2}(p_2)$.

Observe that for case $p_1 = p_2 = 0$, $\widehat{\mathfrak{R}}_1^{U_1}$ and $\widehat{\mathfrak{R}}_1^{U_2}$ are the same set and coincide with that obtained in the normal–normal model.

Bayesian optimal treatments

The marginal expected values of the predictive distribution of c and e are $\bar{c}_i$ and $\bar{e}_i$ (Kotz and Nadarajah, 2004). Thus, under the utility function $U_1(z|R)$ and the penalizing constant p_1, the sets $\mathfrak{R}_1^{U_1}(p_1)$ and $\widehat{\mathfrak{R}}_1^{U_1}(p_1)$ for the Bayesian and the frequentist predictive distributions coincide.

Finally, for the utility $U_2(z|R)$ we proceed by simulation from (4.29) to obtain the Bayesian predictive distribution of the net benefit of the treatment. The next example illustrates this situation.

Example 4.3 (continued). *We consider the artificial data sets in Table 4.2 plus $s_{c_1 e_1} = 2.57$ and $s_{c_2 e_2} = 4.51$.*

We generate a sample of size 10^6 from the Bayesian predictive distribution $\mathcal{T}_2 \left((c, e)^\top | (\bar{c}_i, \bar{e}_i)^\top, \frac{n_i + 1}{n_i - 2} S_i, n_i - 2 \right)$ for $i = 1, 2$. For this sample and a grid of values of R we estimate the probability $\Pr(Z_1 \geq Z_2 | R, \text{data})$ by the proportion of times that z_1 is greater than or equal to z_2 across the simulated samples. Figure 4.7 shows the results.

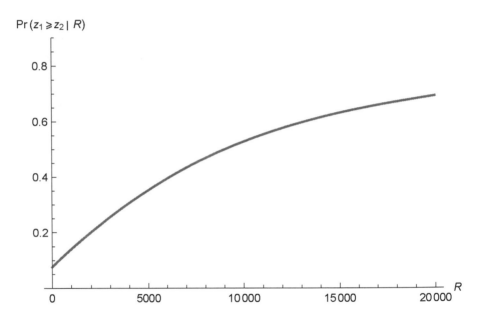

FIGURE 4.7

$\Pr(Z_1 \geq Z_2 | R, \text{data})$ as a function of R in Example 4.3.

If we penalize treatment T_1 by $p_2 = 0.25$, then $\mathfrak{R}_1^{U_2}(0.25) = \{R : R \geq 14559.00\}$.

4.6.5   The dependent lognormal–Bernoulli model

This model is a extension of the lognormal–Bernoulli model given in Section 4.6.3 to the case of parameter dependency. Let (c, e) be the random cost and effectiveness of a treatment T_i such that e follows a Bernoulli distribution $\text{Be}(e|\theta_i)$ and c, conditional on e, follows the lognormal distribution $\Lambda(c|\mu_i(e), \sigma_i(e)^2)$ for $e = 0, 1$. Let us consider the reference prior $\pi_i(\mu_i(e), \sigma_i(e)) \propto 1/\sigma_i(e)$, and the Jeffreys prior $\pi(\theta_i) = \text{Beta}(\theta_i|1/2, 1/2)$.

Frequentist predictive distribution

For a sample $(\mathbf{c}_i, \mathbf{e}_i)$ of treatment T_i, the frequentist predictive distribution of c and e is

$$\widehat{P}_i(c, e|\mathbf{c}_i, \mathbf{e}_i) = \Lambda(c|\tilde{c}_i(e), \tilde{s}_{c_i}^2(e)) \times \text{Be}(e|\bar{e}_i),$$

where

$$n_i(e) = \sum_{j=1}^{n_i} \mathbf{1}_{(e_{ij}=e)}(e), \quad \tilde{c}_i(e) = \frac{1}{n_i(e)} \sum_{j=1}^{n_i} \log(c_{ij}) \mathbf{1}_{(e_{ij}=e)}(e),$$

and

$$\tilde{s}_{c_i}^2(e) = \frac{1}{n_i(e)} \sum_{j=1}^{n_i} (\log(c_{ij}) - \tilde{c}_i(e))^2 \mathbf{1}_{(e_{ij}=e)}(e),$$

for $e = 0, 1$.

Then the frequentist predictive distribution of z renders

$$\widehat{P}_i(z|R, \mathbf{c}_i, \mathbf{e}_i) = (1 - \bar{e}_i) \, \Lambda(-z|\tilde{c}_i(0), \tilde{s}_{c_i}^2(0)) \mathbf{1}_{(-\infty, 0)}(z)$$
$$+ \, \bar{e}_i \, \Lambda(R - z|\tilde{c}_i(1), \tilde{s}_{c_i}^2(1)) \mathbf{1}_{(-\infty, R)}(z). \quad (4.30)$$

Bayesian predictive distribution

The Bayesian predictive net benefit distribution is obtained as in (4.27) and it turns out to be

$$P_i(z|R, n_i, \mathbf{c}_i, \mathbf{e}_i)$$
$$= \frac{n_i(1 - \bar{e}_i) + 0.5}{n_i + 1} \, P_i(-z|n_i(0), \tilde{c}_i(0), \tilde{s}_{c_i}(0)) \mathbf{1}_{(-\infty, 0)}(z)$$
$$+ \frac{n_i \bar{e}_i + 0.5}{n_i + 1} \, P_i(R - z|n_i(1), \tilde{c}_i(1), \tilde{s}_{c_i}(1)) \mathbf{1}_{(-\infty, R)}(z). \quad (4.31)$$

Example 4.8. *Let us consider the dependent lognormal–Bernoulli model and artificial data with $n = 20$, $(n(0), \tilde{c}(0), \tilde{s}_c(0)) = (5, 9.16, 0.06)$ and $(n(1), \tilde{c}(1), \tilde{s}_c(1)) = (15, 9.03, 0.12)$.*

Figure 4.8 displays the Bayesian and frequentist predictive distributions of the net benefit for $R = 10000$. They are bimodal densities, and the Bayesian densities have modes at points -9468.03 and 1777.55, and the frequentist at -9474.82 and 1769.52.

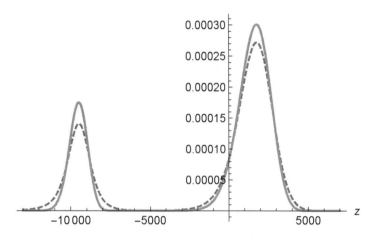

FIGURE 4.8

Bayesian (dashed line) and frequentist (continuous line) predictive distributions of the net benefit for $R = 10000$ in Example 4.8.

Frequentist optimal treatments

For the utility function $U_1(z|R)$ and penalizing constant p_1, treatment T_1 is optimal for any R in the set

$$\widehat{\mathfrak{R}}_1^{U_1}(p_1) = \left\{ R : (\bar{e}_1 - \bar{e}_2)R - (d_1 - d_2) \geq p_1 \right\},$$

where

$$d_i = (1 - \bar{e}_i) \exp\left\{ \tilde{c}_i(0) + \frac{\tilde{s}_{c_i}^2(0)}{2} \right\} + \bar{e}_i \exp\left\{ \tilde{c}_i(1) + \frac{\tilde{s}_{c_i}^2(1)}{2} \right\},$$

for $i = 1, 2$. T_2 is optimal for any R in the set $\widehat{\mathfrak{R}}_2^{U_1}(p_1) = \mathbb{R}^+ - \widehat{\mathfrak{R}}_1^{U_1}(p_1)$.

For the utility function $U_2(z|R)$, the set $\mathfrak{R}_2^{U_2}(p_2)$ is found by simulation from (4.30).

Bayesian optimal treatments

From Lemma 4.2, only the utility function $U_2(z|R)$ can be considered for this model. The set $\mathfrak{R}_1^{U_2}(p_2)$ is found by simulation from (4.31).

Example 4.9. *We consider the data in Table 4.8 from two treatments assuming they are drawn from dependent lognormal–Bernoulli models.*

TABLE 4.8

Data in Example 4.9.

Treatment	$(n(0), \tilde{c}(0), \tilde{s}_c(0))$	$(n(1), \tilde{c}(1), \tilde{s}_c(1))$
T_1	$(5, 9.16, 0.06)$	$(15, 9.03, 0.12)$
T_2	$(10, 9.50, 0.03)$	$(10, 8.90, 0.15)$

The sets of R for which treatment T_1 is optimal for the utility functions $U_1(z|R)$ and $U_2(z|R)$ in an equal–footing scenario are shown in Table 4.9.

TABLE 4.9

$\mathfrak{R}_1^{U_j}$ and $\widehat{\mathfrak{R}}_1^{U_j}$ sets, $j = 1, 2$, for the Bayesian and frequentist predictive distributions and the utility functions $U_1(z|R)$ and $U_2(z|R)$ in Example 4.9.

Utility	Bayesian	Frequentist
U_1	It does not exist	$\widehat{\mathfrak{R}}_1^{U_1} = \mathbb{R}^+$
U_2	$\mathfrak{R}_1^{U_2} = \mathbb{R}^+$	$\widehat{\mathfrak{R}}_1^{U_2} = \{R : R \leq 8907.21\}$

4.7 A case study

This is a cost–effectiveness analysis carried out by Moreno et al. (2010) with real data from a clinical trial in which there are four highly active antiretroviral treatment protocols applied to asymptomatic HIV patients (Pinto et al., 2000). Each treatment protocol combines three drugs chosen from the following five: stavudine (d4T), lamivudine (3TC), indinavir (IND), didanosine (ddl) and azidothymidine (AZT), as follows. Treatment T_1: d4T+3TC+IND, T_2: d4T+ddl+IND, T_3: AZT+3TC+ IND and T_4: AZT+ddl+IND. The effectiveness is described by a dichotomous success–not success random variable, where success means that the patient has no detectable virus load. Effectiveness data are summarized in Table 4.10.

TABLE 4.10

Observed effectiveness in the clinical trial.

Treatment	Success	Not Success	Total
T_1	175	94	269
T_2	51	44	95
T_3	48	43	91
T_4	15	10	25

The collected cost data of the patients in euros are summarized in Table 4.11. There are some large sample values of the cost of the first treatment that suggest a lognormal distribution to accommodate them.

We assume the parametric lognormal–Bernoulli sampling model with the standard reference prior for the parameters of the treatments, and compute the Bayesian optimal decisions.

A partial analysis of the problem that compares T_3 against T_2 and T_3 against T_1 yields the following conclusions. The Bayesian optimal decision under U_2 indicates that T_3 is the optimal one against T_1 and T_2 for all values of the parameter R. For instance, the posterior probability

TABLE 4.11

Observed costs in the clinical trial.

Treatment	Mean	St. Dev.
T_1	7142.28	1568.12
T_2	7302.70	1702.85
T_3	6239.50	931.60
T_4	6282.62	609.18

that treatment T_1 is net benefit larger than T_3 is always less than $1/2$, and in fact it varies in the interval $(0.28, 0.45)$ as R varies.

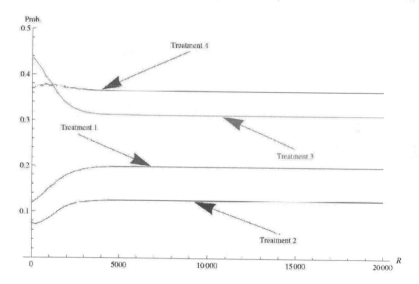

FIGURE 4.9

Multiple treatment comparisons.

On the other hand, a full analysis of the problem should consider multiple comparisons. The optimal decisions with respect to the utility function U_2 are derived from the four curves

$$\left\{ \mathbb{E}_{P_j} U_2(z_j | R) = \Pr\left(Z_j \geq Z_{-j} | R, \text{data} \right), \; j = 1, \ldots, 4 \right\},$$

that we plot in Figure 4.9.

Figure 4.9 indicates that for small values of R, treatment T_3 has the highest expected utility, but for values of R greater than 1100 euros, T_4 has the highest expected utility. We also observe that T_2 has, uniformly on R, the smallest expected utility, and T_1 has the second smallest expected utility. We note that the utility function U_1 cannot be used for the objective Bayesian analysis of the lognormal–Bernoulli reward distributions.

4.8   The cost–effectiveness acceptability curve for the utility function U_1

For the case of two treatments T_1 and T_2 with rewards $P_1(z|R)$ and $P_2(z|R)$, the difference of the expectation of the utility function $U_1(z|R)$ with respect to these rewards yields the well–known incremental net benefit given in equation (1.3) in Chapter 1, and the decision rule is that treatment T_1 is optimal for a value R if

$$\text{INB}_{12}(R) = \mathbb{E}_{P_1}(z|R) - \mathbb{E}_{P_2}(z|R) \geq 0.$$

Thus, the set of R for which treatment T_1 is optimal is

$$\mathfrak{R}_1^{U_1} = \big\{ R : \text{INB}_{12}(R) \geq 0 \big\},$$

and hence T_2 is optimal for any R in

$$\mathfrak{R}_2^{U_1} = \big\{ R : \text{INB}_{12}(R) \leq 0 \big\}.$$

When the distribution $P_1(z|R)$ and $P_2(z|R)$ are not completely known, we have to estimate $\mathfrak{R}_1^{U_1}$ and $\mathfrak{R}_2^{U_1}$.

This is the setting in which appears the cost–effectiveness acceptability curve, and we give a detailed account of this notion when the distribution of the net benefit is completely unknown and when it is known except for a parameter.

Let $\mathbf{e}_i = (e_{i1}, \ldots, e_{in_i})$ and $\mathbf{c}_i = (c_{i1}, \ldots, c_{in_i})$ denote a sample of effectiveness and cost from patients under treatment T_i, $i = 1, 2$. Since the estimate of the optimal treatment, conditional on R, depends on these samples, the question is how the probability $\Pr(\text{INB}_{12}(R) \geq 0)$ behaves as the samples vary in their sample spaces.

We note that for continuous cost and effectiveness,

$$\Pr(\text{INB}_{12}(R) \geq 0) = 1 - \Pr(\text{INB}_{21}(R) \geq 0).$$

Definition 4.1. *The curve*

$$\psi_1(R) = \Pr(INB_{12}(R) \geq 0), \tag{4.32}$$

for $R \geq 0$ is called the cost–effectiveness acceptability curve (CEAC) for treatments T_1 and T_2.

This curve depends not only on the sampling model of the cost and the effectiveness of treatment T_1 and T_2, but also on the procedure we use for estimating $\text{INB}_{12}(R)$.

In the next two sections we consider the CEAC for the nonparametric case where the sampling model for the cost and the effectiveness is completely unknown, and also for the case of specified sampling models with unknown parameters, which is the parametric case. In this latter situation we consider both the case of using a Bayesian approach for eliminating the parameters and the case of using a frequentist approach.

4.8.1 The case of completely unknown rewards

For a given R, let $\mathbf{z}_i = (z_{i1}, \ldots, z_{in_i})$ be the net benefit of the samples of the effectiveness and cost $(\mathbf{c}_i, \mathbf{e}_i)$ from patients under treatment T_i for $i = 1, 2$. A nonparametric estimation of $P_i(z|R)$ is given by the empirical distribution

$$F_i(z|R, \mathbf{z}_i) = \frac{1}{n_i} \sum_{j=1}^{n_i} \mathbf{1}_{(z_{ij} \leq z)}(z)$$

for $z \in \mathbb{R}$. Resampling from the distributions $F_1(z|R, \mathbf{z}_1)$ and $F_2(z|R, \mathbf{z}_2)$ we can compute the mean of each sample (bootstrap sample), and the proportion of bootstrap samples for which the mean from

F_1 is greater than that from F_2. This proportion is the bootstrap estimator of the probability $\Pr(\text{INB}_{12}(R) \geq 0)$.

If we repeat those calculations for a grid of points R we obtain a nonparametric estimator of the curve

$$\psi_1(R) = \Pr(\text{INB}_{12}(R) \geq 0).$$

We note that $\psi_1(R)$ is tied to the utility function $U_1(z|R)$. The meaning of this curve is the sampling probability that treatment T_1 is optimal for the utility function U_1 when sampling from the estimated lotteries $F_1(z|R, \mathbf{z}_1)$ and $F_2(z|R, \mathbf{z}_2)$. From $\psi_1(R)$ we can compute the probability that treatment T_2 is optimal as $\psi_2(R) = 1 - \psi_1(R)$.

The rationale of the bootstrap procedure is that for large sample sizes n_i, $i = 1, 2$, the Law of Large Number yields

$$\mathbb{E}_{F_1}(z|R) - \mathbb{E}_{F_2}(z|R) \approx \frac{1}{n_i} \sum_{j=1}^{n_i} z_{1j}^* - \frac{1}{n_i} \sum_{j=1}^{n_i} z_{2j}^*,$$

where $\{z_{1j}^*, j = 1, \ldots, n_1\}$ and $\{z_{2j}^*, j = 1, \ldots, n_2\}$ are bootstrap samples of sizes n_1 and n_2.

It is clear that the accuracy of the estimator of $\psi_j(R)$ depends on the accuracy the empirical distribution $F_i(z|R, \mathbf{z}_i)$ estimates the true distribution $P_i(z|R)$ for $i = 1, 2$. The Glivenko–Cantelli theorem (Loève, 1963, pp. 20–21) guarantees an accurate estimation for large sample size n_i. Thus, for large sample sizes n_1 and n_2 we have that $\text{INB}_{12}(R) \approx \mathbb{E}_{F_1}(z|R) - \mathbb{E}_{F_2}(z|R)$.

Recall that we are assuming that we know nothing *a priori* about $P_1(z|R)$ and $P_2(z|R)$, and hence we are using the bootstrap procedure for estimating $\psi_1(R)$ and $\psi_2(R)$.

Let us illustrate this curve based on the data in Example 1.1 in Chapter 1.

Example 4.10. *As a continuation of Example 1.1, let us consider the sample of size $n_1 = 270$ of costs and effectiveness of treatment T_1, $(c_{1j}, e_{1j}), j = 1, \ldots, 270$. Thus the observed net benefit of patient j*

under treatment T_1 is $z_{1j} = e_{1j} \times R - c_{1j}$. We construct the empirical distribution

$$F_1(z|R, \mathbf{z}_1) = \frac{1}{270} \sum_{j=1}^{270} \mathbf{1}_{(z_{1j} \leq z)}(z). \qquad (4.33)$$

Analogously for the sample from treatment T_2 of size $n_2 = 95$, the empirical distribution is

$$F_2(z|R, \mathbf{z}_2) = \frac{1}{95} \sum_{j=1}^{95} \mathbf{1}_{(z_{2j} \leq z)}(z). \qquad (4.34)$$

Now, resampling from the above distributions we consider 2000 bootstrap replicates to estimate the probability of a positive incremental net benefit. Figure 4.10 indicates that for $R \leq 21500$ euros, the probability that the INB_{21} is greater than zero is smaller than 0.5 and for $R \geq 21500$ the probability that the INB_{21} is greater than zero is greater than 0.5. We note that this assertion is with respect to the bootstrap samples.

Extension to more than two treatments

It is straightforward to extend the CEAC notion to situations where the number of treatments is greater than 2. Indeed, let $\{T_1, \ldots, T_k\}$, $k \geq 3$, be alternative treatments with completely unknown lotteries $\{P_i(z|R), i = 1, \ldots, k\}$. Let $\psi_j(R)$ denote the probability that treatment T_j is optimal under U_1, conditional on R, that is

$$\psi_j(R) = \Pr\left(\mathbb{E}_{P_j}(z|R) - \max_{i=1,\ldots,k} \mathbb{E}_{P_i}(z|R) = 0\right).$$

For a given R, let $\mathbf{z}_i = (z_{i1}, \ldots, z_{in_i})$ denote a net benefit sample from patients under treatment T_i. Resampling from the empirical distributions $\{F_i(z|R, \mathbf{z}_i), i = 1, \ldots, k\}$ and taking the means of these bootstrap samples, a nonparametric estimation of $\psi_i(R)$ is the proportion of bootstrap samples such that the mean for treatment T_i is greater than the mean for any other treatment. Repeating those calculations for a grid of points R we have a nonparametric estimator of the curve $\psi_i(R)$ for $R \geq 0$.

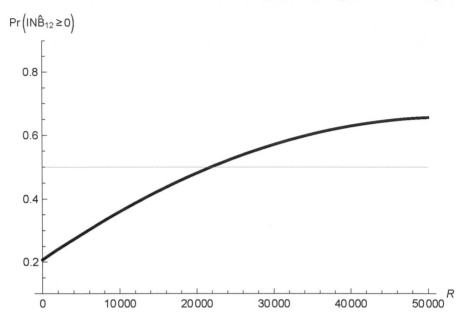

FIGURE 4.10

Cost–effectiveness acceptability curve in Example 1.1 for 2000 bootstrap replications.

4.8.2   The case of parametric rewards

The situation where we know nothing about the model of cost and the effectiveness is not an usual situation in cost–effectiveness analysis. The common practice is that we are able to propose a model for the cost and the effectiveness although the model might depend on some unknown parameters. That is, we start from a specific parametric model $\{P_i(c, e | \theta_i), \ \theta_i \in \Theta\}$ for $i = 1, 2$, where the parameter θ_i is unknown. Then, the incremental net benefit

$$\mathrm{INB}_{12}(R, \theta_1, \theta_2) = \mathbb{E}_{P_1}(z | R, \theta_1) - \mathbb{E}_{P_2}(z | R, \theta_2), \qquad (4.35)$$

depends on the parameters θ_1 and θ_2. For estimating the $\mathrm{INB}_{12}(R, \theta_1, \theta_2)$ we need to estimate θ_1 and θ_2.

Frequentist CEAC

Let (c_{ij}, e_{ij}) for $j = 1, \ldots n_i$ be n_i random variables i.i.d. from $P_i(e, c | \theta_i)$, $\hat{\theta}_i = \hat{\theta}_i(\mathbf{c}_i, \mathbf{e}_i)$ the MLE estimator, and $P_i(c, e | \hat{\theta}_i)$ the predictive distribution for $i = 1, 2$. Then, for each R, the INB_{12} is estimated as the difference of the expectations of these distributions, that is,

$$\widehat{\mathrm{INB}}_{12}(R, \hat{\theta}_1, \hat{\theta}_2) = \mathbb{E}_{P_1}(z|R, \hat{\theta}_1) - \mathbb{E}_{P_2}(z|R, \hat{\theta}_2).$$

The accuracy of this estimator depends on the accuracy of the MLE estimators.

In this parametric setting the cost–effectiveness acceptability curve $\widehat{\psi}_1(R) = \Pr(\widehat{\mathrm{INB}}_{12}(R, \hat{\theta}_1, \hat{\theta}_2) \geq 0)$ is computed with respect to the distribution $g_i(\hat{\theta}_i)$ of the MLE $\hat{\theta}_i(\mathbf{c}_i, \mathbf{e}_i)$ for $i = 1, 2$. That is,

$$\widehat{\psi}_1(R) = \int_I g_1(\hat{\theta}_1) g_2(\hat{\theta}_2) \, d\hat{\theta}_1 d\hat{\theta}_2,$$

where

$$I = \left\{ (\hat{\theta}_1, \hat{\theta}_2) : \widehat{\mathrm{INB}}_{12}(R, \hat{\theta}_1, \hat{\theta}_2) \geq 0 \right\}.$$

In general, it is hard to figure out the distribution $g_j(\hat{\theta}_j)$ so the computation of CEAC is not necessarily an easy task. Further, to identify the set I in the sample space is not, in general, simple. However, for large sample size n_j a normal distribution can be an approximation to the distribution $g_j(\hat{\theta}_j)$, $j = 1, 2$.

Bayesian CEAC

If we complete the sampling model $P_i(c, e|\theta_i)$ by adding a prior for θ_i, we have the Bayesian model $\{P_i(c, e|\theta_i), \pi_i(\theta_i)\}$ for $i = 1, 2$.

Let us consider the Bayesian predictive distribution for the cost and effectiveness of treatment T_i

$$\tilde{P}_i(c, e|\mathbf{c}_i, \mathbf{e}_i) = \int P_i(c, e|\theta_i) \pi_i(\theta_i|\mathbf{c}_i, \mathbf{e}_i) \, d\theta_i,$$

for $i = 1, 2$.

Then, the Bayesian estimation of the $\mathrm{INB}_{12}(R)$ is

$$\widetilde{\mathrm{INB}}_{12}(R, \mathbf{c}_1, \mathbf{e}_1, \mathbf{c}_2, \mathbf{e}_2) = \mathbb{E}_{\tilde{P}_1}(z|R, \mathbf{c}_1, \mathbf{e}_1) - \mathbb{E}_{\tilde{P}_2}(z|R, \mathbf{c}_2, \mathbf{e}_2).$$

In this Bayesian setting the cost–effectiveness acceptability curve $\tilde{\psi}_1(R) = \Pr(\widetilde{\text{INB}}_{12}(R, \mathbf{c}_1, \mathbf{e}_1, \mathbf{c}_2, \mathbf{e}_2) \geq 0)$ is computed with respect to the marginal distribution of the samples $m_i(\mathbf{e}_i, \mathbf{c}_i)$ for $i = 1, 2$, that is

$$\tilde{\psi}_1(R) = \int_{I'} \prod_{i=1}^{2} m_i(\mathbf{c}_i, \mathbf{e}_i) \, d\mathbf{c}_i d\mathbf{e}_i,$$

where

$$I' = \left\{ (\mathbf{c}_1, \mathbf{e}_1, \mathbf{c}_2, \mathbf{e}_2) : \widetilde{\text{INB}}_{12}(R, \mathbf{c}_1, \mathbf{e}_1, \mathbf{c}_2, \mathbf{e}_2) \geq 0 \right\}.$$

To identify the set I' in the sample space is not, in general, an easy task.

The formal extension to the case where we have more than two treatments is immediate and hence it is omitted.

4.9 The cost–effectiveness acceptability curve for the utility function U_2

The above ideas apply word–by–word to the case where the utility function is $U_2(z|R)$ instead of $U_1(z|R)$. In the parametric setting, treatment T_1 is preferred to T_2 for the utility function U_2 if

$$\varphi(R, \theta_1, \theta_2) = \Pr(Z_1 \geq Z_2|R, \theta_1, \theta_2) - \Pr(z_2 \geq z_1|R, \theta_1, \theta_2) \geq 0.$$

These probabilities are computed with respect to the distribution $P_i(c, e|\theta_i)$, for $i = 1, 2$.

Then, if we use the MLE $\hat{\theta}_i = \hat{\theta}_i(\mathbf{c}_i, \mathbf{e}_i)$ for estimating θ_i, we have that the estimation of $\varphi(R, \theta_1, \theta_2)$ is given by

$$\hat{\varphi}(R, \hat{\theta}_1, \hat{\theta}_2) = \Pr(Z_1 \geq Z_2|R, \hat{\theta}_1, \hat{\theta}_2) - \Pr(Z_2 \geq Z_1|R, \hat{\theta}_1, \hat{\theta}_2),$$

where the probabilities have been computed with respect to the predictive distribution $P_i(z_i|\hat{\theta}_i)$ for $i = 1, 2$.

The frequentist cost–effectiveness acceptability curve for the utility function $U_2(z|R)$ is then given by

$$\psi_2(R) = \Pr\left(\widehat{\varphi}(R, \hat{\theta}_1, \hat{\theta}_2) \geq 0\right),$$

where the probability is computed with respect to the sampling distribution of $\hat{\theta}_1(\mathbf{c}_1, \mathbf{e}_1)$ and $\hat{\theta}_2(\mathbf{c}_2, \mathbf{e}_2)$.

On the other hand, using the Bayesian predictive distribution $\tilde{P}_i(c, e|\mathbf{c}_i, \mathbf{e}_i)$ for $i = 1, 2$, the Bayesian estimation of $\varphi(R, \theta_1, \theta_2)$ is given by

$$\widetilde{\varphi}(R, \mathbf{c}_1, \mathbf{e}_1, \mathbf{c}_2, \mathbf{e}_2) = \int_{\Theta \times \Theta} \varphi(R, \theta_1, \theta_2)\pi_1(\theta_1|\mathbf{c}_1, \mathbf{e}_1)\pi_2(\theta_2|\mathbf{c}_2, \mathbf{e}_2)\, d\theta_1 d\theta_2.$$

The Bayesian cost–effectiveness acceptability curve for the utility function $U_2(z|R)$ is then given by

$$\psi_2'(R) = \Pr\left(\widetilde{\varphi}(R, \mathbf{c}_1, \mathbf{e}_1, \mathbf{c}_2, \mathbf{e}_2) \geq 0\right),$$

where the probability is computed with respect to the marginal distribution

$$m(\mathbf{c}_1, \mathbf{e}_1, \mathbf{c}_2, \mathbf{e}_2) = \prod_{i=1}^{2} m_i(\mathbf{c}_i, \mathbf{e}_i).$$

We remark that $\psi_2(R)$ and $\psi_2'(R)$ represent sampling evaluations of the frequentist and Bayesian optimal treatment T_1 under the utility function U_2 when sampling from the true model.

4.10   Comments on cost–effectiveness acceptability curve

The CEAC notion dates back to Van Hout et al. (1994) and Löthgren and Zethraeus (2000), who tried defining the p–value associated with the null hypothesis $H_0 : \text{INB}_{12}(R) \geq 0$ for every R, an interpretation largely adopted in the literature (Briggs and Fenn, 1998; Raikou et al., 1998). We note that the probability that $\text{INB}_{12}(R) \geq 0$ when sampling from a model such that $\text{INB}_{12}(R) \leq 0$ is the type II error, and the probability that $INB_{12}(R) \leq 0$ when sampling from a model such that $INB_{12}(R) \geq 0$ is the type I error. We remark that this testing procedure is not at all an evaluation of the probability of the optimal treatment when sampling from the true model, which is the definition of the CEAC.

Both the frequentist and Bayesian CEACs are sampling evaluations of the procedure we use for estimating the unknown parameters of the lotteries when sampling repeatedly from the true lottery of the cost and the effectiveness. An interpretation of the CEAC for every value R is the proportion of samples from the true model for which the optimal treatment beats the alternative treatments. This notion is certainly quite involved.

5

Cost–effectiveness analysis for heterogeneous data

5.1 Introduction

Let $T_1, \ldots, T_k$ be k alternative treatments for a given disease and the parametric class of distributions of their cost and effectiveness $P(c, e|\theta_1), \ldots, P(c, e|\theta_k)$, where $\theta_1, \ldots, \theta_k$ are unknown parameters.

So far we have assumed that the samples $\mathbf{c}_i = (c_{i1}, \ldots, c_{in_i})$ and $\mathbf{e}_i = (e_{i1}, \ldots, e_{in_i})$ of the cost and effectiveness of treatment T_i are homogeneous, that is, the observations come from a unique distribution $P(c, e|\theta_i)$, $i = 1, \ldots, k$. However, a frequent situation is the one where the data $(\mathbf{c}_i, \mathbf{e}_i)$ are heterogeneous, that is, they are an aggregate of samples

$$\mathbf{c}_i = (\mathbf{c}_{i1}, \ldots, \mathbf{c}_{ih}), \qquad \mathbf{e}_i = (\mathbf{e}_{i1}, \ldots, \mathbf{e}_{ih}),$$

where the subsamples $\mathbf{c}_{ij} = (c_{ij1}, \ldots, c_{ijn_{ij}})$, $\mathbf{e}_{ij} = (e_{ij1}, \ldots, e_{ijn_{ij}})$ are such that $n_i = \sum_{j=1}^{h} n_{ij}$, and come from the distribution $P(c, e|\theta_{ij})$ for $j = 1, \ldots, h$. If full heterogeneity is present we have that $\theta_{ij} \neq \theta_{is}$ for $j \neq s$ and $j, s = 1, \ldots, h$. The homogeneous case corresponds to $\theta_i = \theta_{i1} = \cdots = \theta_{ih}$.

Heterogeneous samples are typically obtained when the subsamples of cost and effectiveness, $\mathbf{c}_{ij}$ and $\mathbf{e}_{ij}$, of treatment T_i are collected in h different health–care centers.

When the samples $(\mathbf{c}_i, \mathbf{e}_i)$ are heterogeneous a central problem is figuring out how to compute the predictive distribution $P(c, e|\mathbf{c}_i, \mathbf{e}_i)$ of the cost and the effectiveness of treatment T_i for $i = 1, \ldots, k$. We note that from the data $(\mathbf{c}_{ij}, \mathbf{e}_{ij})$ we can find $P(c, e|\mathbf{c}_{ij}, \mathbf{e}_{ij})$, the predictive distribution of (c, e) conditional on health center j. However, the quantity of interest is not the predictive distribution conditional on center j, but we are interested in the unconditional predictive distribution $P(c, e|\mathbf{c}_i, \mathbf{e}_i)$ of treatment T_i. The statistical procedure that accounts for the between–sample heterogeneity, and is able to provide the unconditional predictive distribution of (c, e), is called meta–analysis.

Most of the applications of meta–analysis have been carried out in clinical trials, where samples of effectiveness are collected from patients in h different health–care centers, and the quantity of interest is the effectiveness of the treatment. DuMouchel and Waternaux (1992) encouraged the use of meta–analysis in medicine and in controlled clinical trials of psychopharmacological agents by asserting that even when the study protocols are similar (dosage, length of treatment, control treatment), there is often considerable variation between studies. Heterogeneity of the data might also appear in multicenter studies even when the protocols are identical.

The literature on meta–analysis is large and the main statistical model is the so–called *random effect model* (DerSimonian and Laird, 1986; Malec and Sendrask, 1992; Consonni and Veronese, 1995; Evans, 2001; Sutton and Higgins, 2008; Bhaumik et al., 2012; Cornell et al., 2014; Moreno et al., 2014; Friede et al., 2017, among others).

In this chapter we assume the existence of certain heterogeneity degree in the samples of cost and effectiveness across health–care centers, and develop the cost–effectiveness analysis for such heterogeneous data. In this setting, the uncertainty of the underlying models is higher than that of the preceding chapter, where the data were assumed to be homogeneous. Indeed, if the samples $\{(\mathbf{c}_{ij}, \mathbf{e}_{ij}),\ j = 1, \ldots, h\}$ are independent and homogeneous across centers, the sampling model of the whole sample $(\mathbf{c}_i, \mathbf{e}_i)$ is given by

$$P(\mathbf{c}_i, \mathbf{e}_i|\theta_i) = \prod_{j=1}^{h} \prod_{r=1}^{n_{ij}} P(c_{ijr}, e_{ijr}|\theta_i) = \prod_{r=1}^{n_i} P(c_{ir}, e_{ir}|\theta_i). \qquad (5.1)$$

The dimension of this model is the dimension of parameter θ_i. If the samples $\{(\mathbf{c}_{ij}, \mathbf{e}_{ij}), \ j = 1, ..., h\}$ are fully heterogeneous the sampling model is

$$P(\mathbf{c}_i, \mathbf{e}_i | \theta_{i1}, \ldots, \theta_{ih}) = \prod_{j=1}^{h} \prod_{r=1}^{n_{ij}} P(c_{ijr}, e_{ijr} | \theta_{ij}). \qquad (5.2)$$

The model dimension is now the sum of the dimensions of θ_{ij} for $j = 1, \ldots, h$.

It is clear that model (5.2) for full heterogeneity is much more complex than the model for homogeneity (5.1), and whether or not the dimension of model (5.2) can be reduced by grouping those models that have the same parameter is an interesting question that might imply a reduction of the model uncertainty. Therefore, based on the sampling information $\{(\mathbf{c}_{ij}, \mathbf{e}_{ij}), \ j = 1, \ldots, h\}$ we want to know whether some of the parameters $\theta_{i1}, \ldots, \theta_{ih}$ are equal, that is, what centers have the same distribution. We note that we are accepting that the situation of full heterogeneity (5.2) is not necessarily present, and partial heterogeneity is deemed possible. The investigation of the heterogeneity of the data $\{(\mathbf{c}_{ij}, \mathbf{e}_{ij}), \ j = 1, \ldots, h\}$ is a statistical multiple testing problem, a model selection problem which is called *probabilistic clustering*.

Therefore, before a meta–analysis is applied for finding the unconditional predictive distribution of cost and the effectiveness of the treatments, it is convenient to apply a statistical clustering procedure to detect equalities between parameters $\theta_{i1}, \ldots, \theta_{ih}$, for every $i = 1, \ldots, k$. A Bayesian *product partition models* clustering procedure (Hartigan, 1990; Barry and Hartigan, 1992; Casella et al., 2014) will be presented in Section 5.2. After that, in Section 5.3, we will present a general *Bayesian* meta–analysis to account for the resulting heterogeneity of the samples. Then, we will consider in Sections 5.4 and 5.5 the predictive distributions of (c, e) and z. Optimal treatments are computed in Section 5.7, and examples with real data are given in Section 5.8.

Before we undertake this task, we illustrate the consequences of ignoring clustering in the estimation of the treatment effectiveness in the presence of heterogeneity. We give an example with simulated data and use the standard *random effect model.*

The popular version of the random effect model for meta–analysis in clinical trials is the following normal–normal hierarchical model (DerSimonian and Laird, 1986). Let p_j be the probability of success of treatment T conditional on center j for $j = 1, \ldots, h$, and e_j independent effectiveness outcomes from n_j patients receiving the treatment. The distribution of e_j is the binomial

$$\Pr(e_j|p_j, n_j) = \binom{n_j}{e_j} p_j^{e_j}(1 - p_j)^{n_j - e_j}, \ e_j = 0, 1, \ldots, n_j.$$

The quantity of interest is the unconditional effectiveness of treatment T, which is denoted by θ. This parameter θ is known as the *meta–parameter.*

The standard random effect model assumes that the logit transformation of the data

$$y_j = \log\left(\frac{e_j}{n_j - e_j}\right)$$

follows the normal distribution

$$y_j|\theta_j, \sigma_j \sim \mathcal{N}(y_j|\theta_j, \sigma^2),$$

where θ_j is the unknown true treatment effect in center j, and σ^2 the unknown variance.

It is also assumed that the distribution linking the experimental parameter θ_j and the meta–parameter θ is normal

$$\theta_j|\theta, \tau \sim \mathcal{N}(\theta_j|\theta, \tau^2), \ j = 1, \ldots, h,$$

where θ is the true treatment effect, and τ^2 the variance of θ_j. When $\tau = 0$ there is no variation in the treatment effects across centers, and we then have full homogeneity. However, when $\tau > 0$ its meaning is simply the standard deviation of θ_j, and it does not say anything about the variability of θ_j across centers.

Specific formulae for estimating the meta–parameter θ can be seen in DerSimonian and Laird (1986). Some estimated corrections have been considered in Hartung and Knapp (2001) and Bhaumik et al. (2012).

Example 5.1 illustrates the consequences of ignoring clustering in the estimation of the meta–parameter θ.

Example 5.1. *We consider six independent random variables* $\{x_j, j = 1, \ldots, 6\}$ *such that* x_j *follows the Binomial distribution* $Bin(x|n_j, \theta_j)$, *where* $n_1 = \cdots = n_5 = 100$, $n_6 = 500$, *and* $\theta_j = 0.1$ *for* $j = 1, \ldots, 5$ *and* $\theta_6 = 0.7$.

We simulate the data x_j *from the binomial distribution* $Bin(x|\theta_j, n_j)$ *and consider the logit transformation*

$$y_j = \log\left(\frac{x_j}{n_j - x_j}\right)$$

for $j = 1, \ldots, 6$. *The first five samples contain a certain number of zeros, and a continuity correction is applied to the zeros to assure that the logit transformation has meaning.*

Using the random effect model for $h = 6$ *and samples* $\mathbf{y} = (y_1, \ldots, y_6)$, *we compute the estimate* $\hat{\theta}(\mathbf{y})$ *of the meta–parameter* θ. *Then, we repeat the simulation 1000 times and the mean across simulations of the estimator of the meta–parameter* θ *turns out to be 0.18.*

If we use the random effect model after grouping the samples in two groups with samples $\mathbf{s} = (s_1, s_2)$, *where*

$$s_1 = \log\left(\frac{\sum_{j=1}^{5} x_j}{\sum_{j=1}^{5} n_j - \sum_{j=1}^{5} x_j}\right),$$

with $\sum_{j=1}^{5} n_j = 500$ *and* $s_2 = y_6$ *with size 500, the mean of the estimate* $\hat{\theta}(\mathbf{s})$ *of the meta–parameter* θ *across simulations turns out to be 0.35.*

The difference between the two estimates is astonishing. The latter estimate is almost twice the former. The reason for the difference is that the latter estimate incorporates information on how the samples have been clustered before proceeding with the meta–analysis, while the former estimate does not take into account this fact and gives a misleading inference.

5.2   Clustering

As we mentioned above, for a given treatment T_i we want clustering of those models in the set $\{P(c, e|\theta_{ij}), \; j = 1, \ldots, h\}$ that have the same parameter based on the sampling information provided by the data $\{(\mathbf{c}_{ij}, \mathbf{e}_{ij}), \; j = 1, \ldots, h\}$, and we want to do it for treatments $T_1, \ldots, T_k$.

To simplify the notation, in this section we consider a generic treatment T with rewards $\{P(c, e|\theta_j), \; j = 1, \ldots, h\}$ and samples $\{(\mathbf{c}_j, \mathbf{e}_j), j = 1, \ldots, h\}$ from these distributions.

Definition 5.1. *The samples $(\mathbf{c}_j, \mathbf{e}_j)$ from the distribution $P(c, e|\theta_j)$ and $(\mathbf{c}_s, \mathbf{e}_s)$ from $P(c, e|\theta_s)$ are in the same cluster if $\theta_j = \theta_s$.*

If parameters $\{\theta_j, \; j = 1, \ldots, h\}$ are all equal we have only one cluster, and this recovers the homogeneous case. In this case the likelihood of the unique parameter θ for the sample $(\mathbf{c}, \mathbf{e}) = \{(\mathbf{c}_j, \mathbf{e}_j), j = 1, \ldots, h\}$ is given by

$$P(\mathbf{c}, \mathbf{e}|\theta) = \prod_{j=1}^{h} P(\mathbf{c}_j, \mathbf{e}_j|\theta). \tag{5.3}$$

If parameters $\{\theta_j, \; j = 1, \ldots, h\}$ are all different we have h clusters, and every distribution $P(c, e|\theta_j)$ is a cluster. In that case the likelihood of the parameters $\{\theta_j, \; j = 1, \ldots, h\}$ for the sample $(\mathbf{c}, \mathbf{e})$ is

$$P(\mathbf{c}, \mathbf{e}|\theta_1, \ldots, \theta_h) = \prod_{j=1}^{h} P(\mathbf{c}_j, \mathbf{e}_j|\theta_j). \tag{5.4}$$

We note that (5.3) and (5.4) represent two extreme situations of clustering and that intermediate situations, where the number of clusters in the samples is between 1 and h, are certainly possible. Therefore, the problem is to know the cluster structure of the observed data $\{(\mathbf{c}_j, \mathbf{e}_j), \; j = 1, \ldots, h\}$. This is a model selection problem in which the class of models is defined by the class of all possible clusters of the data. We give here the Bayesian approach to this model selection problem following the *product partition models* approach introduced by Hartigan (1990) and further studied by Barry and Hartigan (1992), Crowley (1997) and Casella et al. (2014).

Let us introduce some notation. The sample of cost and effectiveness from health–care center j are simply denoted as $\mathbf{x}_j = (\mathbf{c}_j, \mathbf{e}_j)$, and by $\mathbf{x} = (\mathbf{x}_1, \ldots, \mathbf{x}_h)$ the joint sample. In what follows we refer indistinctly to clustering models $\{P(c, e|\theta_j), j = 1, \ldots, h\}$ or clustering samples $\{\mathbf{x}_j, j = 1, \ldots, h\}$.

The integer p, $1 \leq p \leq h$, denotes the number of clusters we form with the samples $\{\mathbf{x}_j, \ j = 1, \ldots, h\}$. A partition of the sample in p clusters is defined by the vector $\mathbf{r}_p = (r_1, \ldots, r_h)$, where r_j is an integer between 1 and p that indicates the cluster to which $\mathbf{x}_j$ is assigned. For instance, for $h = 4$ and $p = 3$, the partition $\mathbf{r}_3 = (1, 2, 1, 3)$ indicates the clustering $\{\mathbf{x}_1, \mathbf{x}_3\}, \{\mathbf{x}_2\}, \{\mathbf{x}_4\}$.

By $\Re_p$ we denote the set of possible partitions $\mathbf{r}_p$. The number of partitions in $\Re_p$ is the Stirling number of second kind $S(h, p)$. This number is given by

$$S(h, p) = \frac{1}{p!} \sum_{j=0}^{p} (-1)^{p-j} \binom{p}{j} j^h.$$

The set of all possible partitions is $\Re = \cup_{p=1}^{h} \Re_p$, and the number of partitions is given by the Bell number $\mathcal{B}_h = \sum_{p=1}^{h} S(h, p)$.

Given a partition $\mathbf{r}_p$ with p clusters, the sampling model conditional on partition $\mathbf{r}_p$ is given by the product

$$P(\mathbf{x}|p, \mathbf{r}_p, \boldsymbol{\theta}_p, h) = \prod_{j=1}^{p} \prod_{\substack{i:r_i=j \\ i=1,\ldots,h}} P(\mathbf{x}_i|\theta_j), \tag{5.5}$$

where $\boldsymbol{\theta}_p = (\theta_1, \ldots, \theta_p)$ is an unknown parameter in the space Θ^p, and the component θ_j indicates the parameter of the distribution of the sample $\mathbf{x}_i$ such that $r_i = j$.

We remark that the partition $\mathbf{r}_p$ defines the sampling model, so that partition and model are equivalent words in this context.

The partition $\mathbf{r}_1 = (1, \ldots, 1)$ corresponds to the singular case of only one cluster, the homogeneous case, and in this notation the sampling distribution, conditional on this partition, is

$$P(\mathbf{x}|1, \mathbf{r}_1, \theta, h) = \prod_{j=1}^{h} P(\mathbf{x}_j|\theta).$$

It is interesting to note that the singular model $P(\mathbf{x}|1, \mathbf{r}_1, \theta, h)$ is nested into any cluster model $P(\mathbf{x}|p, \mathbf{r}_p, \boldsymbol{\theta}_p, h)$.

5.2.1 Prior distributions

Given the sampling cluster model $P(\mathbf{x}|p, \mathbf{r}_p, \boldsymbol{\theta}_p, h)$, to complete the specification of the Bayesian cluster model, a prior distribution for the discrete parameters $(p, \mathbf{r}_p)$ and for the typically continuous parameter $\boldsymbol{\theta}_p$ have to be specified. A natural decomposition of the prior distribution $\pi(p, \mathbf{r}_p, \boldsymbol{\theta}_p|h)$ is

$$\pi(p, \mathbf{r}_p, \boldsymbol{\theta}_p|h) = \pi(\boldsymbol{\theta}_p|p, \mathbf{r}_p, h)\pi(\mathbf{r}_p|p, h)\pi(p|h).$$

General priors $\pi(\boldsymbol{\theta}_p|p, \mathbf{r}_p, h)$, $\pi(\mathbf{r}_p|p, h)$, and $\pi(p|h)$ have been introduced in Casella et al. (2014). The priors they recommend can be summarized in the three following points.

1. The prior for the parameter θ of the homogeneous model $P(\mathbf{x}|1, \mathbf{r}_1, \theta, h)$ is assumed to be the reference prior $\pi^N(\theta)$. This is the commonly used objective prior for estimating θ (Berger et al., 2009).

    The prior for parameter $\boldsymbol{\theta}_p$ of the cluster model $P(\mathbf{x}|p, \mathbf{r}_p, \boldsymbol{\theta}_p, h)$ is the intrinsic prior arising from the model comparison $P(\mathbf{x}|p, \mathbf{r}_p, \theta_p, h)$ against $P(\mathbf{x}|1, \mathbf{r}_1, \theta, h)$, and it was presented in two steps as in Chapter 2, Section 2.5.1. In a first step, the intrinsic prior for θ_j conditional on θ, is obtained using the intrinsic methodology for the model comparison between model $\{P(\mathbf{y}|\theta_j), \pi^N(\theta_j)\}$ and $P(\mathbf{y}|\theta)$, where $\pi^N(\cdot)$ is the reference prior, θ is an arbitrary but fixed point in Θ, and $\mathbf{y}$ is the training sample of size t such that

$$0 < \int_{\Theta} P(\mathbf{y}|\theta_j)\pi^N(\theta_j)\, d\theta_j < \infty.$$

The dimension t can be fixed as the minimal integer for which these inequalities hold (Berger and Pericchi, 1996a; Moreno et al., 1998). This conditional intrinsic prior for θ_j is given by

$$\pi^I(\theta_j|\theta) = \pi^N(\theta_j) \int_{\mathcal{X}} \frac{P(\mathbf{y}|\theta)}{\int_{\Theta} P(\mathbf{y}|\theta_j)\pi^N(\theta_j)\, d\theta_j} P(\mathbf{y}|\theta_j)\, d\mathbf{y}. \qquad (5.6)$$

Then, the intrinsic prior for $\boldsymbol{\theta}_p$, conditional on θ, is defined as

$$\pi^I(\boldsymbol{\theta}_p|p, \mathbf{r}_p, \theta, h) = \prod_{j=1}^{p} \pi^I(\theta_j|\theta).$$

In a second step, the unconditional intrinsic prior for $\boldsymbol{\theta}_p$ is obtained as

$$\pi^I(\boldsymbol{\theta}_p|p, \mathbf{r}_p, h) = \int_{\Theta} \pi^I(\boldsymbol{\theta}_p|p, \mathbf{r}_p, \theta, h)\pi^N(\theta)\, d\theta.$$

This prior is a useful prior distribution for model selection. Further, it has been applied to different contexts, including cost–effectiveness analysis, change point problems, variable selection in regression, etc. (Berger and Pericchi, 1996a, 2001; Berger et al., 2014; Casella and Moreno, 2006, 2009; Girón et al., 2007; Moreno et al., 2013a,b, 2015, among others).

2. The rationale to define $\pi(\mathbf{r}_p|p, h)$ is as follows. The class of partitions in p clusters $\mathfrak{R}_p$ is first decomposed by noting that if k_i is the number of samples assigned to the ith cluster for $i = 1, \ldots, p$, then the class $\mathfrak{R}_p$ can be expressed as

$$\mathfrak{R}_p = \bigcup_{\substack{1 \le k_1 \le \cdots \le k_p \\ k_1 + \cdots + k_p = h}} \mathfrak{R}_{p;k_1,\ldots,k_p},$$

where $\mathfrak{R}_{p;k_1,\ldots,k_p}$ is the class of partitions in $\mathfrak{R}_p$ having sizes $(k_1, \ldots, k_p)$. To the vector $(k_1, \ldots, k_p)$ we call a *configuration*.

Using this decomposition of $\mathfrak{R}_p$ we can now write the prior $\pi(\mathbf{r}_p|p, h)$ as

$$\pi(\mathbf{r}_p|p, h) = \pi(\mathbf{r}_p|\mathfrak{R}_{p;k_1,\ldots,k_p}, h)\pi(\mathfrak{R}_{p;k_1,\ldots,k_p}|p, h).$$

Since the labels of the clusters are irrelevant, the number of partitions in $\mathfrak{R}_{p;k_1,\ldots,k_p}$ is given by

$$\binom{h}{k_1\cdots k_p}\frac{1}{R(k_1,\ldots k_p)},$$

where $\binom{h}{k_1\cdots k_p}$ is the multinomial coefficient, and

$$R(k_1,\ldots,k_p) = \prod_{i=1}^{h}\left(\sum_{j=1}^{p}\mathbf{1}_{(k_j=i)}\right)!$$

corrects the count by considering the redundant strings in the vector $(k_1,\ldots,k_p)$. For instance, for a vector $(k_1,\ldots,k_p)$ such that $k_1 = \cdots = k_{p-4} < k_{p-3} = k_{p-2} < k_{p-1} = k_p$, we have that $R(k_1,\ldots,k_p) = (p-4)!2!2!$.

Since the partitions $\mathbf{r}_p$ in $\mathfrak{R}_{p;k_1,\ldots,k_p}$ are exchangeable, it seems reasonable to assign a uniform prior to them, that is

$$\pi(\mathbf{r}_p|\mathfrak{R}_{p;k_1,\ldots,k_p}, h) = \binom{h}{k_1\cdots k_p}^{-1}R(k_1,\ldots,k_p).$$

Further, since the configuration classes

$$\left\{\mathfrak{R}_{p;k_1,\ldots,k_p},\ 1 \le k_1 \le \cdots \le k_p,\ k_1 + \cdots + k_p = h\right\}$$

in $\mathfrak{R}_p$ contain models of the same complexity, it seems reasonable to assign to these classes a uniform prior. Therefore, to define this uniform prior we only need to count the number of configuration classes in $\mathfrak{R}_p$. We note that this number is also the number of ways the integer h can be partitioned into p ordered integer parts, which we denote by $b(h, p)$. This number does not seem to have a closed form expression as a function of p and h. However, it can be shown that $b(h, p)$ satisfies the recursive equation

$$b(h, p) = b(h - 1, p - 1) + b(h - p, p), \ 1 \le p \le h,$$

with the restrictions

$$b(h, 1) = b(h, h) = 1.$$

These recursive equations can be numerically solved. Therefore, we have

$$\pi(\Re_{p;k_1,\ldots,k_p} | p, h) = \frac{1}{b(h, p)}, \qquad \Re_{p;k_1,\ldots,k_p} \in \Re_p.$$

3. In the cost–effectiveness analysis scenario there is no reason to penalize, *a priori*, a large number of clusters, so that the prior on the number of clusters $\pi(p|h)$ is assumed to be the uniform distribution

$$\pi(p|h) = \frac{1}{h}, \qquad p = 1, \ldots, h.$$

Thus, we finally have the prior on the discrete parameter $p, \mathbf{r}_p$ as

$$\pi(p, \mathbf{r}_p | h) = \pi(\mathbf{r}_p | p, h)\pi(p|h). \tag{5.7}$$

For instance, for $\mathbf{r}_3 = (1, 2, 1, 3)$ we have that $k_1 = 1$, $k_2 = 1$, $k_3 = 2$, $b(4, 3) = 1$, and hence

$$\pi(\mathbf{r}_3 | p = 3, h = 4) = \frac{3!}{1!1!2!} \frac{1}{2!} \frac{1}{4} = \frac{3}{8}.$$

5.2.2 Posterior distribution of the cluster models

Let
$$M_{\mathbf{r}_p} : \{P(\mathbf{x}|p, \mathbf{r}_p, \boldsymbol{\theta}_p, h), \ \pi(\boldsymbol{\theta}_p | p, \mathbf{r}_p, h)\pi(p, \mathbf{r}_p|h)\},$$

be a Bayesian cluster model defined by the partition $\mathbf{r}_p$ in the class of cluster models with at most h clusters. From the Bayes theorem the posterior probability of $M_{\mathbf{r}_p}$, or equivalently the posterior distribution of the partition $(p, \mathbf{r}_p)$, is given by

$$\Pr(p, \mathbf{r}_p|\mathbf{x}, h) = \frac{m(\mathbf{x}|p, \mathbf{r}_p, h)\pi(p, \mathbf{r}_p|h)}{\sum_{p=1}^{h} \sum_{\mathbf{r}_p \in \Re_p} m(\mathbf{x}|p, \mathbf{r}_p, h)\pi(p, \mathbf{r}_p|h)}, \ M_{\mathbf{r}_p} \in \Re, \ (5.8)$$

where $m(x|p, \mathbf{r}_p, h)$, the marginal of the data, is

$$m(\mathbf{x}|p, \mathbf{r}_p, h) = \int_{\Theta^p} f(\mathbf{x}|p, \mathbf{r}_p, \boldsymbol{\theta}_p, h) \pi^I(\boldsymbol{\theta}_p|p, \mathbf{r}_p, h) \, d\boldsymbol{\theta}_p.$$

Each model $M_{\mathbf{r}_p} \in \mathfrak{R}$ indicates a different heterogeneity structure of the samples $\mathbf{x} = \{\mathbf{x}_i, \, i = 1, \dots, h\}$, and the set of probability distributions $\{\Pr(M_{\mathbf{r}_p}|x), \, M_{\mathbf{r}_p} \in \mathfrak{R}\}$ gives us a measure of the uncertainty we have on these models.

5.2.3 Examples

Clinical trials often include binary data. In this section we use simulated and real binary data to illustrate the accuracy of the clustering procedure for estimating the true cluster model.

Example 5.1 (continued). *As in Example 5.1 we consider six independent samples x_j, $j = 1, \dots, 6$, with sample size $n_1 = \dots = n_5 = 100$ and $n_6 = 500$. The samples are simulated from six Binomial distributions with parameters $\theta_j = 0.1$ for $j = 1, \dots, 5$, and $\theta_6 = 0.7$. Thus, the true model is the cluster model formed by the groups $\{1, 2, 3, 4, 5\}, \{6\}$.*

For $h = 6$ we have 203 cluster models and for the six simulated samples we compute their posterior probabilities. We repeat the simulation 500 times and find the mean of the posterior model probabilities across the simulations. We found that the true model had the largest posterior probability in 485 out of 500 simulations.

In Table 5.1 we present the mean across simulations of the posterior probability of the top cluster models. We note that the mean of the posterior probability of the six cluster model $\{1\}, \{2\}, \{3\}, \{4\}, \{5\}, \{6\}$ is as small as 0.005.

To illustrate the performance of the proposed Bayesian clustering procedure for estimating the true cluster model, when the difference between θ_6 and $\theta_1 = \dots = \theta_5$ decreases, a simulation study has been conducted. Keeping as a true model the clusters $\{1, 2, 3, 4, 5\}, \{6\}$ we consider different values of θ_6 (from 0.2 to 0.7) capturing scenarios from more to less "closeness between clusters." Also, in order to illustrate the behavior of the proposed model with respect to the sample size, we

TABLE 5.1

Mean of the posterior probabilities of the top cluster models across the 500 simulations.

Top cluster model	Mean of the posterior probability
$\{1,2,3,4,5\},\{6\}$ (true model)	0.87
$\{1\},\{2,3,4,5\},\{6\}$	0.015
the rest	< 0.01
other models	
$\{1\},\{2\},\{3\},\{4\},\{5\},\{6\}$ (full heterogeneity)	0.005
$\{1,2,3,4,5,6\}$ (homogeneity)	$< 10^{-4}$

consider the following cases: I: $n_1 = \cdots = n_5 = 10$, $n_6 = 50$; II: $n_1 = \cdots = n_5 = 25$, $n_6 = 125$; III: $n_1 = \cdots = n_5 = 50$, $n_6 = 250$, and IV: $n_1 = \cdots = n_5 = 100$, $n_6 = 500$. Each case is identified by n_6 and thus we refer to it by the corresponding n_6 value.

For all cases, a simulation run consisted of 500 samples. Figure 5.1 displays the estimated posterior probabilities of the true cluster model. Figure 5.1 also shows that when the sample size increases and the distance between clusters increases, both the posterior probabilities of the true model and the average rate of success also increase. This behavior shows a very reasonable perform of the Bayesian product partition model clustering procedure. For instance, from the bottom panel in Figure 5.1 we observe that even for small values of θ_6, the model performs very well for moderately large sample sizes reaching values of the average rate of success above 60%.

Example 5.2. *Data in Table 5.2 correspond to a classical six major multicenter clinical trials carried out in the seventies to assess the beneficial effect of a daily dose of aspirin in post–myocardial infarction patients (Canner, 1987). The data were collected from six major randomized multicenter clinical trials of aspirin and placebo during the period 1970–79 in post–myocardial infarction patients. The trials*

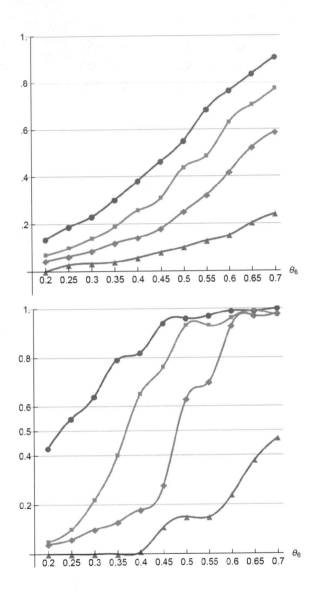

FIGURE 5.1

Top panel: Behavior of the mean of the posterior probabilities of the true model across the 500 simulations for different values of θ_6 from 0.2 to 0.7 by increments of 0.05. **Bottom panel:** Average rate of success of the true model across the 500 simulations. In both pictures, symbols $\triangle$, $\diamond$, $\square$ and $\circ$ refer to $n_6 = 50, 125, 250$, and 500, respectively.

were the First United Kingdom trial (number 1), the Coronary Drug
Project Aspirin trial (number 2), the German–Austrian Multicenter
Study (number 3), the Second United Kingdom trial (number 4), the
Persantine–Aspirin Reinfarction study (number 5), and the Aspirin
Myocardial Infarction Study (number 6).

TABLE 5.2

Number of patients and number of deaths under aspirin and placebo
in six clinical trials.

Trial number	Aspirin		Placebo	
	n_i	x_i	n_i	x_i
1	615	49	624	67
2	758	44	771	64
3	317	27	309	32
4	832	102	850	126
5	810	85	406	52
6	2267	246	2257	219

Using the clustering methodology explained above for binomial sampling models, the resulting cluster models under aspirin and placebo and
their posterior probabilities are displayed in Table 5.3.

We note that the class of cluster models contains 203 models.

TABLE 5.3

Top cluster models for aspirin and placebo data and their posterior
probabilities.

Aspirin		Placebo	
Cluster Model	Post. Prob.	Cluster Model	Post. Prob.
{1,2,3},{4,5,6}	0.17	{1,2,3,6},{4,5}	0.14
{1},{2},{3},{4},{5},{6}	0.09	{1,2,3,5,6},{4}	0.10
{1,2},{3,4,5,6}	0.07	{1},{2},{3},{4},{5},{6}	0.06
{2},{1,3,4,5,6}	0.05	{1,2,3,6},{4},{5}	0.05
the rest	< 0.05	the rest	< 0.05

We observe that the one–cluster model does not appear as one of the top cluster models because the homogeneity has a very small posterior probability.

Example 5.3. *The dataset in Table 5.4 is extracted from Cosmi et al. (2008) and corresponds to the comparison of total mortality in 4 studies for patients treated with ticlopidine (treatment T_1) versus oral antico-agulation for coronary stenting (treatment T_2).*

TABLE 5.4

Number of patients and number of deaths under treatment T_1 and T_2 in Example 5.3.

Study	T_1		T_2	
	n_i	x_i	n_i	x_i
1	243	2	230	4
2	257	1	260	2
3	177	3	173	2
4	546	0	550	0

Table 5.5 shows the top clusters we found from this dataset among the 15 possible cluster models.

TABLE 5.5

Top cluster models for treatment T_1 and T_2 and their posterior probabilities in Example 5.3.

T_1		T_2	
Cluster Model	Post. Prob.	Cluster Model	Post. Prob.
{1,3},{2,4}	0.21	{1,2,3},{4}	0.27
{1,2,4},{3}	0.17	{1,3},{2,4}	0.14
{1,2,3},{4}	0.13	{1},{2,3},{4}	0.09
{1,2,3,4}	0.13	{1,3},{2},{4}	0.08
{1},{2,4},{3}	0.08	{1},{2,3,4}	0.08
{1,2},{3}, {4}	0.07	{1,2,3,4}	0.07
the rest	< 0.05	the rest	< 0.07

We note that the full heterogeneity cluster denoted by $\{1\}, \{2\}, \{3\}, \{4\}$ has a posterior probability lower than 0.03 for any of the treatments, and the homogeneous cluster model $\{1, 2, 3, 4\}$ appears in treatment T_2 with a probability as small as 0.07.

Examples 5.2 and 5.3 illustrate that full heterogeneity or homogeneity are not usual cluster models in multicenter studies.

5.3 Bayesian meta–analysis

In real meta–analysis applications, the true cluster model has to be estimated, and the uncertainty of the model estimation must be incorporated into the meta–inference. The standard random effect model does not take into account the model uncertainty, as it only considers the full heterogeneity model. This, however, can be misleading as illustrated in Example 5.1. The Bayesian meta–analysis model that we introduce in this section does not share this inconvenience and the uncertainty on the cluster model estimation is automatically incorporated into the meta–inference.

Let T be a generic treatment with rewards $\{P(c, e|\theta_j), \; j = 1, \ldots, h\}$, and data $\mathbf{x} = (\mathbf{x}_1, \ldots, \mathbf{x}_h)$, where $\mathbf{x}_j = (\mathbf{c}_j, \mathbf{e}_j)$ is the data from center j for $j = 1, \ldots, h$. The sampling distribution of $\mathbf{x}$, conditional on the partition $\mathbf{r}_p$ of $\mathbf{x}$ in p clusters, is given by

$$P(\mathbf{x}|p, \mathbf{r}_p, \boldsymbol{\theta}_p, h) = \prod_{j=1}^{p} P(\mathbf{y}_j|p, \mathbf{r}_p, \theta_j, h),$$

where

$$P(\mathbf{y}_j|p, \mathbf{r}_p, \theta_j, h) = \prod_{\substack{i:r_i=j \\ i=1,\ldots,n_i}} P(\mathbf{x}_i|\theta_j).$$

We complete the model with the reference prior $\pi^N(\theta_j)$ and the objective Bayesian cluster model $M_{\mathbf{r}_p}$ is the pair

$$M_{\mathbf{r}_p} : \left\{ P(\mathbf{x}|p, \mathbf{r}_p, \boldsymbol{\theta}_p, h), \prod_{j=1}^{p} \pi^N(\theta_j) \right\}. \tag{5.9}$$

5.3.1 The Bayesian meta–model

The cost and the effectiveness $x = (c, e)$ of a generic treatment T are latent nonobservable random variables, and its distribution $P(x|\theta)$ is called the meta–model and θ the meta–parameter. We are only able to observe $x = (c, e)$ in a center, and by $P(x|\theta_j)$ we denote the experimental distribution of x conditional on center j and the experimental parameter θ_j, $j = 1, \ldots, h$.

In the heterogeneous case the meta–model is in the same family as the experimental models $P(x|\theta_j)$, and the prior of the meta–parameter $\pi(\theta)$ is of the same type as that of the experimental model $\pi(\theta_j)$. This prior can certainly be the objective reference prior. Thus, the Bayesian meta–model M is defined by

$$M : \left\{ P(x|\theta), \pi^N(\theta) \right\}. \tag{5.10}$$

5.3.2 The likelihood of the meta–parameter and the linking distribution

Let $\mathbf{x} = \{\mathbf{x}_j, \ j = 1, \ldots, h\}$ be a multiple sample drawn from $\{P(x|\theta_j), j = 1, \ldots, h\}$, such that $\theta_j \neq \theta_s$, for $j \neq s$. Assuming that the data from different centers are independent, conditional on $\theta_1, \ldots, \theta_h$, the likelihood of $\theta_1, \ldots, \theta_h$ for the data $\mathbf{x}$ is given by

$$P(\mathbf{x}|\theta_1, \ldots, \theta_h) = \prod_{j=1}^{h} P(\mathbf{x}_j|\theta_j).$$

With the elements we have so far we cannot find the likelihood of the meta–parameter θ for the sample $\mathbf{x}$. To be able to do that, we have to introduce a prior distribution linking the experimental parameter θ_j and the meta–parameter θ. This linking distribution $\pi(\theta_j|\theta)$ is a probability distribution of θ_j conditional on θ for $j = 1, \ldots, h$.

Assuming that $\theta_1, \ldots, \theta_h$ are *a priori* independent, conditional on θ, the likelihood of the meta–parameter θ for the whole sample $\mathbf{x}$ follows from the likelihood of $\theta_1, \ldots, \theta_h$ and the linking distributions

$\pi(\theta_1|\theta), \ldots, \pi(\theta_h|\theta)$, and it turns out to be

$$P(\mathbf{x}|\theta) \;=\; \int_{\Theta} \prod_{j=1}^{h} P(\mathbf{x}_j|\theta_j)\pi(\theta_j|\theta) \, d\theta_j = \prod_{j=1}^{h} \int_{\Theta} P(\mathbf{x}_j|\theta_j)\pi(\theta_j|\theta) \, d\theta_j$$

$$= \prod_{j=1}^{h} P(\mathbf{x}_j|\theta).$$

Thus, the likelihood of the meta-parameter θ for the multiple sample $\mathbf{x}$ is the product of the likelihoods of θ for the samples $\mathbf{x}_j$ for $j = 1, \ldots, h$.

5.3.3 Properties of the linking distribution

We note that the linking distribution $\pi(\theta_j|\theta)$ for $j = 1, \ldots, h$ cannot be arbitrarily chosen but it has to satisfy the following properties.

1. The linking distribution $\pi(\theta_j|\theta)$ has to be a proper prior, that is, for any θ the equality

$$\int_{\Theta} \pi(\theta_j|\theta) \, d\theta_j = 1,$$

   must be satisfied. Otherwise, the conditional and marginal prior distribution might be incompatible, as will be shown later on.

2. The linking distribution $\pi(\theta_j|\theta)$ has to be compatible with the specified priors $\pi(\theta_j)$ and $\pi(\theta)$. This means that the bivariate distribution

$$\pi(\theta_j, \theta) \;=\; \pi(\theta_j|\theta)\pi(\theta)$$

   must satisfy the integral equations

$$\int_{\Theta} \pi(\theta_j, \theta) \, d\theta_j = \pi(\theta), \qquad \int_{\Theta} \pi(\theta_j, \theta) \, d\theta = \pi(\theta_j). \tag{5.11}$$

   When this occurs, it is said that $\pi(\theta_j, \theta)$ belongs to the Frèchet class of bidimensional distributions with given marginal $\pi(\theta_j)$ and $\pi(\theta)$.

A general linking distribution $\pi(\theta_j|\theta)$ is provided by the conditional intrinsic priors for model selection (Berger and Pericchi, 1996a; Moreno, 1997; Moreno et al., 1998) given in expression (5.6). Let $\pi^I(\theta_j|\theta, t)$ be the conditional distribution arising from the model comparison between the meta–model $M : P(x|\theta)$ for an arbitrary but fixed value θ, and the Bayesian experimental model $M_j : \{P(x|\theta_j), \pi(\theta_j)\}$ for a training sample of size t. This conditional intrinsic prior for a fixed t, is given by

$$\pi^I(\theta_j|\theta, t) = \pi(\theta_j) \int_{\mathbb{R}^t} \frac{\left(\prod_{i=1}^{t} P(x_i|\theta)\right) \left(\prod_{i=1}^{t} P(x_i|\theta_j)\right)}{\int_{\Theta} \left(\prod_{i=1}^{t} P(x_i|\theta_j)\right) \pi(\theta_j)\, d\theta_j}\, dx_1 \ldots dx_t.$$

The conditional intrinsic prior $\pi^I(\theta_j|\theta, t)$ is a probability density for any integer t. For,

$$\int_{\Theta} \pi^I(\theta_j|\theta, t)\, d\theta_j = \int_{\Theta} \pi(\theta_j)\, d\theta_j \int_{\mathbb{R}^t} \frac{\left(\prod_{i=1}^{t} P(x_i|\theta)\right) \left(\prod_{i=1}^{t} P(x_i|\theta_j)\right)}{\int_{\Theta} \left(\prod_{i=1}^{t} P(x_i|\theta_j)\right) \pi(\theta_j)\, d\theta_j}$$

$$dx_1 \ldots dx_t.$$

Reversing the order of integration in the right side of this equation we have

$$\int_{\Theta} \pi^I(\theta_j|\theta, t)\, d\theta_j = \int_{\mathbb{R}^t} \left(\prod_{i=1}^{t} P(x_i|\theta)\right)$$

$$\times \frac{\int_{\Theta} \left(\prod_{i=1}^{t} P(x_i|\theta_j)\right) \pi(\theta_j) d\theta_j}{\int_{\Theta} \left(\prod_{i=1}^{t} P(x_i|\theta_j)\right) \pi(\theta_j)\, d\theta_j}\, dx_1 \ldots dx_t$$

$$= 1.$$

We assume that parameters $\theta_1, \ldots, \theta_p$ are independent, conditional on θ and t, and hence the conditional distribution of $\boldsymbol{\theta}_p = (\theta_1, \ldots, \theta_p)$ is given by

$$\pi(\boldsymbol{\theta}_p|\theta, t) = \prod_{j=1}^{p} \pi^I(\theta_j|\theta, t). \tag{5.12}$$

Lemma 5.1 shows that the intrinsic joint distribution $\pi^I(\theta_j, \theta|t)$ satisfies equations (5.11).

Lemma 5.1. *The bidimensional distribution*

$$\pi^I(\theta_j, \theta|t) = \pi^I(\theta_j|\theta, t)\pi(\theta)$$

has marginals $\pi(\theta)$ *and* $\pi(\theta_j)$ *for any integer* t.

Proof. We first note that for any t we have

$$\int_\Theta \pi^I(\theta_j, \theta|t) \, d\theta_j = \pi(\theta) \int_\Theta \pi^I(\theta_j|\theta, t) \, d\theta_j = \pi(\theta)$$

and hence the first equation in (5.11) is satisfied.

On the other hand, for any t we have

$$\int_\Theta \pi^I(\theta_j, \theta|t) \, d\theta = \pi(\theta_j) \times$$

$$\int_{\mathbb{R}^t} \frac{\int_\Theta \left(\prod_{i=1}^t P(x_i|\theta) \right) \pi(\theta) \, d\theta}{\int_\Theta \left(\prod_{i=1}^t P(x_i|\theta) \right) \pi(\theta_j) \, d\theta_j} \prod_{i=1}^t P(x_i|\theta_j) \, dx_i.$$

Since $\prod_{i=1}^t P(x_i|\theta) = P(x_1, \ldots, x_t|\theta)$ and $\prod_{i=1}^t P(x_i|\theta_j) = P(x_1, \ldots, x_t|\theta_j)$ belong to the same family of parametric distributions, we have that

$$\frac{\int_\Theta P(x_1, \ldots, x_t|\theta)\pi(\theta) \, d\theta}{\int_\Theta P(x_1, \ldots, x_t|\theta_j)\pi(\theta_j) \, d\theta_j} = 1.$$

Therefore,

$$\int_\Theta \pi^I(\theta_j, \theta|t) \, d\theta = \pi(\theta_j) \int_{\mathbb{R}^t} P(x_1, \ldots, x_t|\theta_j) \, dx_1 \ldots dx_t$$

$$= \pi(\theta_j),$$

and this proves the assertion. $\qquad\qquad\square$

The training sample size t controls the concentration degree of the probability distribution of θ_j around θ. Lemma 5.2 shows that, under mild conditions, as t tends to infinity, the distribution $\pi^I(\theta_j|\theta, t)$ degenerates to a point mass on θ. In other words, the larger the sample size t, the larger the concentration of the conditional intrinsic prior around the meta–parameter.

Lemma 5.2. *For any regular sampling model $P(x|\theta)$ we have that $\pi^I(\theta_j|\theta, t)$ degenerates to a point mass on θ as t tends to infinity. Further, if $\lim_{t\to\infty} \pi^I(\theta_j|\theta, t)$ is a probability density we then have that*

$$\lim_{t\to\infty} \pi^I(\theta_j|\theta, t) = \delta_{\{\theta\}}(\theta_j),$$

where $\delta_{\{\theta\}}(\theta_j)$ represents the Dirac delta.

Proof. We note that $\pi^I(\theta_j|\theta, t)$ can be written as

$$\pi^I(\theta_j|\theta, t) = \pi(\theta_j)\mathbb{E}_{x_1,\ldots,x_t|\theta_j} B_{01}(x_1,\ldots,x_t),$$

where $B_{01}(x_1,\ldots,x_k)$ is the Bayes factor to compare model

$$M_0 : P(x|\theta), \text{ for fixed } \theta,$$

versus model

$$M_1 : \{P(x|\theta_j), \pi(\theta_j)\},$$

for the sample $x_1,\ldots,x_t$. Using the consistency property of Bayes factors for nested models (Casella *et al.* 2009), it follows that the limit in probability when sampling from model M_1 is zero, that is

$$\lim_{t\to\infty} B_{01}(x_1,\ldots,x_t) = 0, \quad [P_{\theta_j}],$$

and hence $\mathbb{E}_{x_1,\ldots,x_t|\theta_j} B_{01}(x_1,\ldots,x_t)$ goes to zero as $t \to \infty$. Thus, the distribution $\pi^I(\theta_j|\theta, t)$ degenerates to zero when $t \to \infty$ and $\theta_j \neq \theta$.

Further, when $\lim_{t\to\infty} \pi^I(\theta_j|\theta, t)$ is a probability density, it follows that

$$\lim_{t\to\infty} \pi^I(\theta_j|\theta, t) = \begin{cases} 0, & \text{for } \theta_j \neq \theta, \\ 1, & \text{for } \theta_j = \theta, \end{cases}$$

and this completes the proof of Lemma 5.2.   □

5.3.4   Examples

We give two examples to illustrate the linking distribution for Bernoulli and normal Bayesian models.

Example 5.4. *Let $\{e_j, \ j = 1, \ldots, h\}$ be the independent random effectiveness of a treatment in h health centers. We suppose that e_j follows a Bernoulli distribution with parameter θ_j. The prior for θ_j is the Jeffreys prior, $\pi(\theta_j) = Beta(\theta_j | 1/2, 1/2)$. Hence, the Bayesian experimental models are*

$$M_j : \left\{ Be(e_j | \theta_j), Beta(\theta_j | 1/2, 1/2) \right\}, \quad j = 1, \ldots, h,$$

and then, the Bayesian meta–model is

$$M : \{ Be(e | \theta), Beta(\theta | 1/2, 1/2) \} .$$

Then, the intrinsic prior for θ_j, conditional on θ and t, turns out to be

$$\pi^I(\theta_j | \theta, t) = \sum_{i=0}^{t} Beta(\theta_j | i + 1/2, t - i + 1/2) \, Bin(i | t, \theta), \qquad (5.13)$$

a Beta–Binomial mixture distribution. In Figure 5.2 we plot the distribution $\pi^I(\theta_j | \theta, t)$ for $\theta = 0.4$ and several values of t. We note that the larger the t, the higher the concentration of the intrinsic distribution around 0.4.

From Lemma 5.1 it follows that the marginals of the joint distribution $\pi^I(\theta_j, \theta | t) = \pi^I(\theta_j | \theta, t) \, Beta(\theta | 1/2, 1/2)$ are the Jeffreys prior for θ_j and θ for any integer t. Thus the conditional and marginal distributions of (θ_j, θ) are coherent.

It is interesting to note that the correlation coefficient between θ_j and θ is $\rho = t/(t + 2)$, and as t varies over the integer set, the correlation coefficient varies over the interval $(1/3, 1)$. Thus, a wide range of correlation between θ_j and θ can be modeled as the hyperparameter t varies.

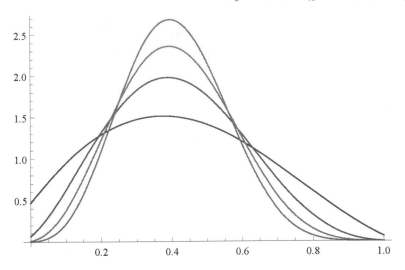

FIGURE 5.2

From bottom to top intrinsic linking distributions for $\theta = 0.4$ and $t = 5, 10, 15$ and 20.

Example 5.5. *Let c_j, $j = 1, \ldots, h$, be the independent random cost of a treatment such that c_j follows a normal distribution with mean and variance (θ_j, σ_j^2), $j = 1, \ldots, h$. The reference prior for (θ_j, σ_j) is $\pi^N(\theta_j, \sigma_j) = k/\sigma_j \mathbf{1}_{\mathbb{R} \times \mathbb{R}+}(\theta_j, \sigma_j)$, where k is an arbitrary positive constant. Thus, the objective experimental cost Bayesian models are*

$$M_j : \{\mathcal{N}(c_j | \theta_j, \sigma_j^2), \pi^N(\theta_j, \sigma_j)\}, \quad j = 1, \ldots, h.$$

Then, the Bayesian meta–model is

$$M : \{\mathcal{N}(c | \theta, \sigma^2), \pi^N(\theta, \sigma)\},$$

where the meta–parameters θ and σ^2 are the mean and variance of the cost of the treatment.

The intrinsic prior for θ_j, σ_j, conditional on the minimal training sample size $t = 2$ and the meta–parameters (θ, σ), is given in Section 2.5.1 Example 2.12 as

$$\begin{aligned}
\pi^I(\theta_j, \sigma_j | \theta, \sigma) &= \pi(\theta_j | \theta, \sigma, \sigma_j) \pi(\sigma_j | \sigma) \\
&= \mathcal{N}\left(\theta_j | \theta, \frac{1}{2}(\sigma^2 + \sigma_j^2)\right) HC^+(\sigma_j | \sigma), \quad (5.14)
\end{aligned}$$

where $HC^+(\sigma_j|\sigma)$ denotes the half Cauchy distribution

$$HC^+(\sigma_j|\sigma) = \frac{2}{\pi\sigma(1 + \sigma_j^2/\sigma^2)}.$$

From Lemma 5.1 it follows that the joint distribution $\pi^I(\theta_j, \sigma_j, \theta, \sigma)$ $= \pi^I(\theta_j, \sigma_j|\theta, \sigma)\pi^N(\theta, \sigma)$ has marginals $\pi^N(\theta_j, \sigma_j)$ and $\pi^N(\theta, \sigma)$, and hence the distributions $\pi^I(\theta_j, \sigma_j|\theta, \sigma)$, $\pi^N(\theta_j, \sigma_j)$ and $\pi^N(\theta, \sigma)$ are coherent.

Example 5.5 (continued). *For the Bayesian model*

$$M_j : \{\mathcal{N}(c_j|\theta_j, \sigma_j^2), \ \pi^N(\theta_j, \sigma_j)\}$$

and meta–model

$$M : \{\mathcal{N}(c|\theta, \sigma^2), \ \pi^N(\theta, \sigma)\},$$

an alternative choice to the intrinsic linking distribution $\pi^I(\theta_j, \sigma_j|\theta, \sigma)$ is the following normal–inverse–gamma distribution,

$$\pi(\theta_j, \sigma_j|\theta, \sigma) = \mathcal{N}(\theta_j|\theta, \sigma_j^2) \ IG\left(\sigma_j|1/2, \sigma^2/2\right), \qquad (5.15)$$

where

$$IG(\sigma_j|1/2, \sigma^2/2) = \left(\frac{2}{\pi}\right)^{1/2} \frac{\sigma^{1/2}}{\sigma_j^2} \exp\left(-\frac{\sigma^2}{2\sigma_j^2}\right),$$

is the inverse–gamma distribution with parameter $(1/2, \sigma^2/2)$.

Next lemma proves that the linking distribution $\pi(\theta_j, \sigma_j|\theta, \sigma)$ in (5.15) and the marginals $\pi^N(\theta_j, \sigma_j)$ and $\pi^N(\theta, \sigma)$ are coherent.

Lemma 5.3. *The joint distribution*

$$\pi(\theta_j, \sigma_j, \theta, \sigma) = \mathcal{N}(\theta_j|\theta, \sigma_j^2) \ IG(\sigma_j|1/2, \sigma^2/2)\frac{k}{\sigma},$$

has marginals $\pi^N(\theta_j, \sigma_j)$ and $\pi^N(\theta, \sigma)$.

Proof. The first assertion follows from

$$\int_0^\infty \int_{-\infty}^\infty \pi(\theta_j, \sigma_j | \theta, \sigma) \pi^N(\theta, \sigma) \, d\theta d\sigma = \int_0^\infty IG(\sigma_j | 1/2, \sigma^2/2) \frac{k}{\sigma} \, d\sigma$$

$$\int_{-\infty}^\infty \mathcal{N}(\theta_j | \theta, \sigma_j^2) \, d\theta$$

$$= \int_0^\infty IG(\sigma_j | 1/2, \sigma^2/2) \frac{k}{\sigma} \, d\sigma$$

$$= \pi^N(\theta_j, \sigma_j).$$

The second assertion follows from

$$\int_0^\infty \int_{-\infty}^\infty \pi(\theta_j, \sigma_j | \theta, \sigma) \pi^N(\theta, \sigma) \, d\theta_j d\sigma_j$$

$$= \frac{k}{\sigma} \int_0^\infty IG(\sigma_j | 1/2, \sigma^2/2) \, d\sigma_j \int_{-\infty}^\infty \mathcal{N}(\theta_j | \theta, \sigma_j^2) \, d\theta_j = \pi^N(\theta, \sigma).$$

This completes the proof.   □

5.4   The predictive distribution of (c, e) conditional on a partition

Let $\mathbf{r}_p$ be a partition in p clusters of the data $\mathbf{x} = (\mathbf{x}_1, \ldots, \mathbf{x}_h)$. The Bayesian model is that given in (5.9), that is,

$$M_{\mathbf{r}_p} : \left\{ P(\mathbf{x} | p, \mathbf{r}_p, \boldsymbol{\theta}_p, h), \prod_{j=1}^p \pi^N(\theta_j) \right\},$$

and the linking distribution $\pi(\boldsymbol{\theta}_p | \theta) = \prod_{j=1}^p \pi^I(\theta_j | \theta)$. The sampling distribution of $\mathbf{x}$, conditional on the partition $\mathbf{r}_p$ and the meta–parameter

θ, is given by

$$P(\mathbf{x}|p, \mathbf{r}_p, \theta, h) = \int_{\Theta^p} \left(\prod_{j=1}^{p} P(\mathbf{y}_j|p, \mathbf{r}_p, \theta_j, h) \right) \left(\prod_{j=1}^{p} \pi^I(\theta_j|\theta) \right) d\theta_1 \dots d\theta_p$$

$$= \prod_{j=1}^{p} \int_{\Theta} P(\mathbf{y}_j|p, \mathbf{r}_p, \theta_j, h) \pi^I(\theta_j|\theta) \, d\theta_j$$

$$= \prod_{j=1}^{p} P(\mathbf{y}_j|p, \mathbf{r}_p, \theta, h).$$

Then, from the Bayes theorem it follows that the posterior distribution of the meta–parameter θ, conditional on the partition $\mathbf{r}_p$, is given by

$$\pi(\theta|\mathbf{x}, p, \mathbf{r}_p, h) = \frac{P(\mathbf{x}|p, \mathbf{r}_p, \theta, h)\pi^N(\theta)}{m(\mathbf{x})}, \tag{5.16}$$

where the marginal of the data $m(\mathbf{x})$ is

$$m(\mathbf{x}) = \int_{\Theta} P(\mathbf{x}|p, \mathbf{r}_p, 0, h)\pi^N(\theta) \, d\theta.$$

We note that the posterior distribution of θ is well defined even when $\pi^N(\theta, \sigma)$ is an improper prior, because the arbitrary constant c cancels out in the ratio of the posterior distribution.

Then, using the meta model of x in (5.10) $M : \{P(x|\theta), \pi^N(\theta)\}$, the Bayesian predictive distribution of the cost and the effectiveness of treatment T, conditional on $\mathbf{r}_p$, is given by

$$P(x|\mathbf{x}, p, \mathbf{r}_p, h) = \int_{\Theta} P(x|\theta)\pi(\theta|\mathbf{x}, p, \mathbf{r}_p, h) \, d\theta. \tag{5.17}$$

An alternative to the Bayesian predictive distribution, conditional on $\mathbf{r}_p$, is the frequentist predictive distribution of the treatment. If we consider the maximum likelihood estimator of $\hat{\theta}_{\mathbf{r}_p} = \hat{\theta}(\mathbf{x}|\mathbf{r}_p)$ of the meta–parameter θ, conditional on $\mathbf{r}_p$, that is,

$$\hat{\theta}(\mathbf{x}) = \arg\sup_{\theta \in \Theta} \ P(\mathbf{x}|p, \mathbf{r}_p, \theta, h),$$

the frequentist predictive distribution of x for treatment T is given by

$$\widehat{P}(x|\mathbf{x}, p, \mathbf{r}_p, h) = P(x|p, \mathbf{r}_p, \hat{\theta}, h). \tag{5.18}$$

5.4.1   The unconditional predictive distribution of (c, e)

Using the Bayesian predictive density $P(c, e | \mathbf{x}, p, \mathbf{r}_p, h)$, conditional on $\mathbf{r}_p$, given in (5.17) and the posterior distribution of the partitions

$$\{\Pr(p, \mathbf{r}_p | \mathbf{x}, h), M_{\mathbf{r}_p} \in \mathfrak{R}\}$$

in (5.8), the unconditional Bayesian predictive distribution of (c, e) is obtained as the mixture

$$P(c, e | \mathbf{x}, h) = \sum_{p=1}^{h} \sum_{\mathbf{r}_p \in \mathfrak{R}_p} P(c, e | \mathbf{x}, p, \mathbf{r}_p, h) \Pr(p, \mathbf{r}_p | \mathbf{x}, h). \qquad (5.19)$$

If we use the frequentist predictive distribution, conditional on $\mathbf{r}_p$, in (5.18) the unconditional frequentist predictive distribution of (c, e) is then given by

$$\widehat{P}(c, e | \mathbf{x}, h) = \sum_{p=1}^{h} \sum_{\mathbf{r}_p \in \mathfrak{R}_p} \widehat{P}(c, e | \mathbf{x}, p, \mathbf{r}_p, h) \Pr(p, \mathbf{r}_p | \mathbf{x}, h). \qquad (5.20)$$

We note that the number of terms in the mixtures (5.19) and (5.20) is the Bell number $\mathcal{B}_h$, a huge number even for moderate values of h. Thus, the computation of $P(c, e | \mathbf{x}, h)$ for moderately large values of h is not feasible. However, in clustering applications, most of the posterior probabilities $\{\Pr(p, \mathbf{r}_p | \mathbf{x}, h), \ \mathbf{r}_p \in \mathfrak{R}\}$ used to be negligible, and the number of nonnegligible models becomes small.

5.5   The predictive distribution of the net benefit z

Using the Bayesian predictive distribution of (c, e) in (5.19), the Bayesian predictive net benefit distribution of the treatment for a given

R, unconditional on the cluster structure of the data, is obtained as

$$
\begin{aligned}
P(z|R, \mathbf{x}, h) &= \int_{\mathcal{E}} P(eR - z, e|\mathbf{x}, h) \, de \\
&= \sum_{p=1}^{h} \sum_{\mathbf{r}_p \in \Re_p} \Pr(p, \mathbf{r}_p|\mathbf{x}, h) \int_{\mathcal{E}} P(eR - z, e|p, \mathbf{r}_p, h) de.
\end{aligned}
$$

$$(5.21)$$

If we use the frequentist predictive distribution of (c, e) in (5.20), the frequentist predictive net benefit distribution of the treatment for a given R, unconditional on the cluster structure of the data, is obtained as

$$
\begin{aligned}
\widehat{P}(z|R, \mathbf{x}, h) &= \int_{\mathcal{E}} \widehat{P}(eR - z, e|\mathbf{x}, h) \, de \\
&= \sum_{p=1}^{h} \sum_{\mathbf{r}_p \in \Re_p} \Pr(p, \mathbf{r}_p|\mathbf{x}, h) \int_{\mathcal{E}} \widehat{P}(eR - z, e|p, \mathbf{r}_p, h) \, de.
\end{aligned}
$$

$$(5.22)$$

5.6 The case of independent c and e

When c and e are independent, the Bayesian clustering procedure is applied to the cost samples $\mathbf{c} = \{\mathbf{c}_j, \ j = 1, \ldots, h\}$ and independently to the effectiveness samples $\mathbf{e} = \{\mathbf{e}_j, \ j = 1, \ldots, h\}$, and hence two sets of cluster models and their corresponding posterior distributions are obtained. Let $\{\Pr(p, \mathbf{r}_p|\mathbf{c}, h), \ \mathbf{r}_p \in \Re\}$ and $\{\Pr(p, \mathbf{r}_p|\mathbf{e}, h), \ \mathbf{r}_p \in \Re\}$ be the posterior probabilities of the cluster models.

If we factorize the distribution of (c, e) as $P(c, e|\xi) = P(c|\xi)P(e|\xi)$, where ξ is an unknown parameter, then from the application of the clustering and meta–analysis procedure to the data $\mathbf{c}$, we obtain the Bayesian predictive distribution of the cost as

$$
P(c|\mathbf{c}, h) = \sum_{p=1}^{h} \sum_{\mathbf{r}_p \in \Re_p} P(c|p, \mathbf{r}_p, h) \Pr(p, \mathbf{r}_p|\mathbf{c}, h),
$$

and similarly the Bayesian predictive distribution of the effectiveness as

$$P(e|\mathbf{e}, h) = \sum_{p=1}^{h} \sum_{\mathbf{r}_p \in \Re_p} P(e|p, \mathbf{r}_p, h) \Pr(p, \mathbf{r}_p|\mathbf{e}, h).$$

Therefore, the Bayesian predictive net benefit distribution of the treatment z turns out to be

$$
\begin{aligned}
P(z|R, \mathbf{c}, \mathbf{e}) &= \int_{\mathcal{E}} P(eR - z|\mathbf{c}, h) P(e|\mathbf{e}, h) \, de = \\
&= \int_{\mathcal{E}} \left(\sum_{p=1}^{h} \sum_{\mathbf{r}_p \in \Re_p} P(eR - z|p, \mathbf{r}_p, h) \Pr(p, \mathbf{r}_p|\mathbf{c}, h) \right) \\
&\quad \left(\sum_{p=1}^{h} \sum_{\mathbf{r}_p \in \Re_p} P(e|p, \mathbf{r}_p, h) \Pr(p, \mathbf{r}_p|\mathbf{e}, h) \right) de. \quad (5.23)
\end{aligned}
$$

It is apparent that expression (5.23) cannot be derived from expression (5.21) unless the equality

$$\Pr(p, \mathbf{r}_p|\mathbf{c}, h) = \Pr(p, \mathbf{r}_p|\mathbf{e}, h)$$

holds for any $\mathbf{r}_p \in \Re$. If the equality holds, then (5.23) becomes

$$
\begin{aligned}
P(z|R, \mathbf{c}, \mathbf{e}) &= \int_{\mathcal{E}} P(eR - z|\mathbf{c}, h) P(e|\mathbf{e}, h) \, de \\
&= \sum_{p=1}^{h} \sum_{\mathbf{r}_p \in \Re_p} \Pr(p, \mathbf{r}_p|\mathbf{c}, \mathbf{e}, h) \\
&\quad \int_{\mathcal{E}} P(eR - z|p, \mathbf{r}_p, h) P(e|p, \mathbf{r}_p, h) \, de
\end{aligned}
$$

which is indeed the particular case of (5.21) under the independence assumption.

A similar discussion can be made when using the frequentist predictive distribution instead of the Bayesian predictive distribution.

5.7 Optimal treatments

Let $T_1, \ldots, T_k$ be our set of competitive treatments with reward $P(c, e|\theta_1), \ldots, P(c, e|\theta_k)$, and samples $\mathbf{x}_1, \ldots, \mathbf{x}_k$. Each sample $\mathbf{x}_i$ is an aggregate of samples from h hospitals. Applying formula (5.17) or (5.18) to the data $\mathbf{x}_i$ we get either the unconditional Bayesian predictive distribution of the net benefit $P(z|R, \mathbf{x}_i, h)$ or the unconditional frequentist predictive distribution $\widehat{P}(z|R, \mathbf{x}_i, h)$ of treatment T_i for $i = 1, \ldots, k$.

Using $P(z|R, \mathbf{x}_i, h)$ for $i = 1, \ldots, k$ and a utility function $U(z|R)$, treatment T_j is optimal for any R in the set

$$\mathfrak{R}_j^U = \left\{ R : \mathbb{E}_P(z|R, \mathbf{x}_j, h) = \max_{i=1,\ldots,k} \mathbb{E}_P(z|R, \mathbf{x}_i, h) \right\},$$

where

$$\mathbb{E}_P(z|R, \mathbf{x}_i, h) = \int P(z|R, \mathbf{x}_i, h) U(z|R) \, dz.$$

We have that

$$\mathbb{R}^+ = \cup_{i=1}^k \mathfrak{R}_i^U,$$

where some sets in $\{\mathfrak{R}_j^{U_1}, \ j = 1, \ldots, k\}$ might be empty.

If we use $\widehat{P}(z|R, \mathbf{x}_i, h)$ for $i = 1, \ldots, k$, treatment T_j is optimal for any R in the set

$$\widehat{\mathfrak{R}}_j^U = \left\{ R : \mathbb{E}_{\widehat{P}}(z|R, \mathbf{x}_j, h) = \max_{i=1,\ldots,k} \mathbb{E}_{\widehat{P}}(z|R, \mathbf{x}_i, h) \right\},$$

where

$$\mathbb{E}_{\widehat{P}}(z|R, \mathbf{x}_i, h) = \int \widehat{P}(z|R, \mathbf{x}_i, h) U(z|R) \, dz.$$

Again we have that $\mathbb{R}^+ = \cup_{i=1}^k \widehat{\mathfrak{R}}_i^U$, and some sets in $\{\widehat{\mathfrak{R}}_j^U, \ j = 1, \ldots, k\}$ might be empty.

5.8 Examples

Two examples with real data sets are presented in this section. It is assumed that the distribution of the cost and the effectiveness in each center is normal, and they are independent random variables. All the expectations and probabilities in the examples are conditional not only to R but to the data.

Example 5.6. *The data were obtained from a study to compare three methadone maintenance programs: high, medium and low intensity, for opioid–addicted patients (Puigdollers et al., 2003). A 12–month follow–up study of 586 patients beginning methadone treatment at four drug care centers in Barcelona was performed. The Nottingham Health Profile (NHP) was used to measure quality of life. This is a questionnaire of 38 items that measures quality of life through a scale that varies from 0, the normal health state, to 100, the worst health state.*

The difference between the NHP value at the start of the treatment and the value one month later was used as the effectiveness measure. Although three methadone maintenance programs were compared in the original study, for simplicity of this illustration we only compare the medium–intensity program, to be denoted in the following as treatment T_1, and the high–intensity program, to be denoted as treatment T_2.

To treatment T_1 were randomly assigned 165 patients, and 155 to treatment T_2. Both treatments were applied in four hospitals. Table 5.6 shows the mean and standard deviation of the effectiveness and cost of this data set.

To these data sets we applied a cluster analysis and Table 5.7 displays the top cluster models for the original data from the four hospitals. This analysis shows that the effectiveness and cost data present a strong partially heterogeneous behavior for both treatments. The probability of the one–cluster model $\{1,2,3,4\}$ is close to zero for any of the four data sets.

The cost–effectiveness analysis of treatment T_1 and T_2 for the utility function U_1 for the heterogeneous data is summarized by the continuous straight line in Figure 5.3. This line displays the expectation $\mathbb{E}(z_1 - z_2 | R, data)$ as a function of R for $0 < R < 1000$ euros, and the conclusion is that treatment T_1 is the optimal treatment for $0 < R < 122.7$ euros, and for $122.7 < R < 1000$ the optimal one is treatment T_2.

If the heterogeneity of the data is not considered and we assume that the data are homogeneous, the cost–effectiveness analysis is that

TABLE 5.6

Summary of effectiveness, cost and sample sizes in Example 5.6.

Center	Effectiveness	Cost	Sample size
	Treatment T_1		
Overall	16.59 (20.73)	552.81 (306.33)	126
1	16.61 (21.10)	507.88 (230.98)	23
2	16.73 (20.88)	562.06 (304.17)	39
3	16.81 (21.29)	508.69 (223.29)	29
4	16.91 (20.97)	564.34 (282.47)	35
	Treatment T_2		
Overall	19.10 (21.77)	652.84 (336.38)	133
1	19.33 (21.79)	612.68 (264.18)	34
2	20.43 (21.76)	689.88 (277.62)	35
3	16.11 (21.85)	657.30 (333.69)	32
4	19.33 (21.84)	649.04 (279.85)	32

given by the dashed straight line in Figure 5.3. We note that in this latter case, T_1 is an optimal treatment for the range $0 < R < 36.8$. Therefore, for $36.8 < R < 122.7$ the homogeneous cost–effectiveness analysis chooses treatment T_2 while the heterogeneous cost–effectiveness analysis chooses treatment T_1.

The heterogeneous cost–effectiveness analysis of treatment T_1 and T_2 for the utility function U_2 is summarized by the continuous line in Figure 5.4. This line displays the probability $\Pr(z_1 > z_2|R, data)$, as a function of R euros. We recall that treatment T_1 is now preferred to T_2 if $\Pr(z_1 > z_2|R, data) \geq 0.5$. The conclusion is that treatment T_1 is the optimal treatment for $0 < R < 122.7$ euros, while treatment T_2 is optimal for $122.7 < R < 1000$ euros.

This conclusion does not coincide with the one we would obtain if homogeneity of the data from the four hospitals were assumed. In this latter case, the dashed line in Figure 5.4 shows that treatment T_1 is preferred to T_2 only for $0 < R < 36.8$.

TABLE 5.7

Top clusters for effectiveness and cost data for treatments T_1 and T_2 and their posterior probabilities.

Treatment T_1		Treatment T_2	
Effectiveness Clusters	Posterior Probabilities	Effectiveness Clusters	Posterior Probabilities
$\{1, 3, 4\}, \{2\}$	0.63	$\{1, 3, 4\}, \{2\}$	0.64
$\{1, 3\}, \{2, 4\}$	0.19	$\{1, 3\}, \{2, 4\}$	0.17
$\{1\}, \{2, 3, 4\}$	0.07	$\{1, 4\}, \{2, 3\}$	0.05
the rest	< 0.05	the rest	< 0.05
Cost Clusters	Posterior Probabilities	Cost Clusters	Posterior Probabilities
$\{1, 3\}, \{2, 4\}$	0.86	$\{1\}, \{2, 3, 4\}$	0.52
$\{1\}, \{2\}, \{3\}, \{4\}$	0.12	$\{1, 3\}, \{2, 4\}$	0.17
the rest	< 0.05	the rest	< 0.05

In Example 5.7, we consider the clinical trials conducted by Burns et al. (1999) in four health–care centers for comparing intensive intervention with standard intervention for psychotic patients. The principal effectiveness outcome was the time in hospital for psychiatric problems over 24 months. As part of the baseline data, information was collected on the number of days each patient spent in hospital in the 24 months. The societal cost for each individual over 24 months was estimated by Nixon and Thompson (2005) who evaluated the cost of the treatments according to recorded resources used (including social, hospital and community components).

The effectiveness and cost of each treatment are assumed to be independent. The distribution for the effectiveness in each center is assumed to be normal, and the distribution of the cost lognormal. As we know, this latter assumption implies that the expected Bayesian net benefit does not exist and hence we use the criteria of choosing treatment T_1 if $\Pr(z_1 \geq z_2 | R, \text{data}) \geq 0.5$.

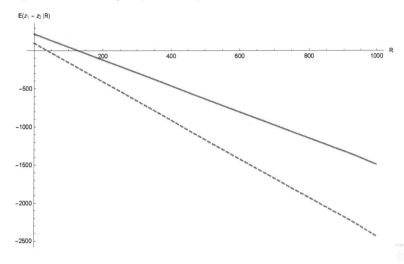

FIGURE 5.3

$\mathbb{E}(z_1 - z_2 | R, \text{data})$ for heterogeneous data (continuous line), and under the assumption that the data are homogeneous (dashed line).

Example 5.7. *Burns et al. (1999) randomly assigned 708 psychotic patients in four centers to standard case management (355 patients) or intensive case management (353 patients). The trial compared intensive intervention (treatment T_2) in which managers had a case load of 10–15 patients, with the standard intervention (treatment T_1), with case loads of 30–35 patients. Table 5.8 shows the mean and standard deviation of the observed cost and effectiveness in Example 5.7.*

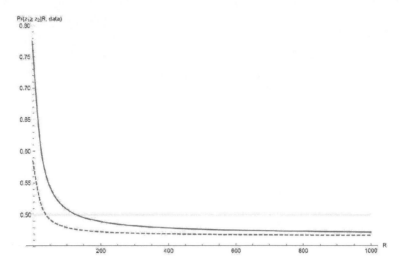

FIGURE 5.4

$\Pr(z_1 > z_2 | R, \text{data})$ for heterogeneous data (continuous line), and under the assumption that the data are homogeneous (dashed line).

The rationale for the trial was that, by providing more intensive support, the patients' use of health and other services might be reduced, and clinical outcomes might be improved. The effectiveness analysis by Burns et al. (1999) led the authors to the conclusion that "There was no significant decline in overall hospital use among intensive–case–management patients."

A clustering analysis of the data reveals the cluster structures we display in Table 5.9, which clearly shows that the data are heterogeneous.

The heterogeneous cost–effectiveness analysis of treatment T_1 and T_2 for the utility function U_2 is summarized in Figure 5.5, where the solid line displays the probability $\Pr(z_1 > z_2 | R, \text{data})$, as a function of $R \in (0, 5000)$ euros.

TABLE 5.8

Summary of effectiveness, cost and sample sizes in Example 5.7.

Center	Effectiveness	Cost	Sample size
	Treatment T_1		
Overall	74.66 (112.45)	22704.19 (22000.33)	332
1	64.27 (85.09)	19434.63 (22743.37)	92
2	60.58 (114.19)	25194.08 (23491.82)	73
3	91.95 (125.24)	22934.61 (19881.67)	98
4	78.86 (121.95)	24102.09 (22190.83)	69
	Treatment T_2		
Overall	74.19 (124.52)	24553.40 (23407.54)	335
1	73.45 (138.44)	24161.33 (25036.03)	93
2	79.92 (129.85)	25455.22 (24772.52)	76
3	66.67 (112.49)	23346.90 (19776.63)	91
4	78.43 (116.40)	25589.62 (24334.89)	75

This solid line shows that for $R \leq 889$ euros, treatment T_1 is the optimal one, while for $R \geq 889$, the optimal treatment is T_2.

Assuming that the data are homogeneous, the probability $\Pr(z_1 > z_2 | R, data)$ as a function of R is given by the dashed line in Figure 5.5. The cost–effectiveness conclusion would then be that treatment T_1 is the optimal one for any R.

TABLE 5.9

Top clusters for cost and effectiveness data for two treatments and their posterior probabilities.

Treatment T_1		Treatment T_2	
Effectiveness Clusters	Posterior Probabilities	Effectiveness Clusters	Posterior Probabilities
$\{1\}, \{2,3,4\}$	0.90	$\{1,2\}, \{3,4\}$	0.27
$\{1,2\}, \{3,4\}$	0.08	$\{1\}, \{2,3,4\}$	0.22
the rest	< 0.05	$\{1,2,4\}, \{3\}$	0.20
		$\{1,2,3\}, \{4\}$	0.13
		$\{1,3,4\}, \{2\}$	0.10
		the rest	< 0.05
Cost Clusters	Posterior Probabilities	Cost Clusters	Posterior Probabilities
$\{1,2\}, \{3,4\}$	0.68	$\{1,2,3\}, \{4\}$	0.23
$\{1\}, \{2,3,4\}$	0.19	$\{1,3,4\}, \{2\}$	0.22
$\{1,2,3\}, \{4\}$	0.11	$\{1\}, \{2,3,4\}$	0.15
the rest	< 0.05	$\{1,2,4\}, \{3\}$	0.14
		$\{1,4\}, \{2,3\}$	0.10
		$\{1,3\}, \{2,4\}$	0.09
		$\{1,2\}, \{3,4\}$	0.08
		the rest	< 0.05

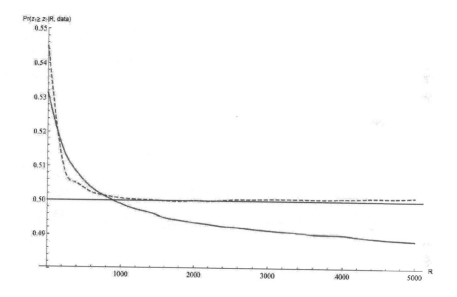

FIGURE 5.5

$\Pr(z_1 \geq z_2 | R, \text{data})$ as a function of R for heterogeneous data (continuous line), and under the assumption that the data are homogeneous (dashed line).

6

Subgroup cost–effectiveness analysis

6.1 Introduction

In preceding chapters we have seen that, given a set of alternative treatments $\{T_1, \ldots, T_k\}$ and the parametric class of distributions of their cost and effectiveness $\{P_i(c, e|\theta_i), i = 1, \ldots, k\}$, the reward of choosing treatment T_i is the predictive distribution of the net benefit $P_i(z|R, \text{data}_i)$, conditional on the amount of money the decision maker is willing to pay for the unit of effectiveness R and the cost and effectiveness data$_i$. Furthermore, if a utility function $U(z|R)$ of the net benefit z is assumed, the cost–effectiveness analysis provides an optimal treatment for the whole patient population, conditional on R and the data sets $\{\text{data}_i, i = 1, \ldots, k\}$.

In this chapter we enlarge the above scheme by adding to the data of cost and effectiveness of every patient, a set of patient covariates $\mathbf{x} = (x_1, \ldots, x_p)$. The covariates indicate certain deterministic physical characteristics of the patient such as age, sex, health status, and semiological variables of the disease (Sculpher and Gafni, 2001; NICE, 2013; Espinoza et al., 2014). In this setting the predictive distribution of the net benefit of treatment T_i depends not only on R and data$_i$ but also on the covariates $\mathbf{x}$, that is, we now have the rewards $P_i(z|R, \text{data}_i, \mathbf{x})$ for $i = 1, \ldots, k$. The adaptation of the cost–effectiveness analysis to this situation yields the cost–effectiveness analysis for subgroups. Thus, the aim of the subgroup cost–effectiveness analysis is that of finding optimal treatments with respect to a given utility function for subgroups of patients defined by $\mathbf{x}$.

A subgroup is formed with those patients sharing specified values of some covariates of $\mathbf{x}$. For instance, a possible subgroup is formed with those patients having an age in a given interval of years and the same sex. It is clear that for a given utility function $U(z|R)$, the expected utility of the reward $P_i(z|R, \text{data}_i, \mathbf{x})$,

$$\mathbb{E}_{P_i}(U) = \int_{\mathcal{Z}} U(z|R)\, dP_i(z|R, \text{data}_i, \mathbf{x}),$$

is a function of R, data_i, and $\mathbf{x}$, for $i = 1, \ldots, k$, and hence the resulting optimal treatment might change as $\mathbf{x}$ changes, so that different subgroups might have different optimal treatments. The optimal treatment for the whole patient population might be suboptimal for a given subgroup.

Since the definition of patient subgroups is made in terms of the set of covariates, it is important to exclude those covariates that do not have an influence on the disease. This means that for carrying out a cost–effectiveness analysis for subgroups, an initial step should be the statistical detection of the influential covariates from the original set of them. The statistical selection of covariates is known as the variable selection problem, an old and important problem in regression analysis for which different methods have been utilized through its long history and include methods based on p–values (R^2, R^2 corrected, Mallows C_p), on Bayes factor approximations (Akaike AIC, Schwarz BIC, and Spiegelhalter et al. DIC), and on genuine Bayes factors for g–priors and intrinsic priors.

We note that the statistical variable selection typically yields not only a more realistic definition of patient subgroups, but also a desirable reduction of the dimension of the regression model. We also note that the statistical variable selection introduces model uncertainty in the cost–effectiveness analysis, an uncertainty that the Bayesian methodology incorporates into the decision making in an automatic way.

A coherent Bayesian solution to the variable selection problem formulates the problem as a particular case of the general model selection problem, and in this chapter, Bayesian variable selection procedures in normal and probit regression models, and some of their statistical properties, are discussed.

On the other hand, in Chapter 4 two utility functions of the net benefit of a given treatment were introduced, the linear utility function $U_1(z|R)$ and the nonlinear utility function $U_2(z|R)$. The former assumes that transfers of health among the patient population are acceptable, and the latter does not. In this chapter we study optimal treatments for subgroups for both the utility function $U_1(z|R)$ and $U_2(z|R)$.

The chapter is organized as follows. The data and the linear models for the cost and the effectiveness of the treatments are described in Section 6.2. In Section 6.3 we present Bayesian statistical procedures for variable selection in normal and probit linear models. For simplicity in the presentation we consider the case where the cost, conditional on the effectiveness, and the effectiveness can be written as a linear model. The particular case where the cost and the effectiveness are independent is also presented.

For the linear model for the selected influential covariates, we compute the Bayesian predictive distribution of the net benefit of the treatments in Section 6.4. In Section 6.5 we present the optimal treatments for subgroups, and in Section 6.6 we illustrate the analysis on examples for simulated and real data. Section 6.7 calls attention to the fact that sometimes the statistical variable selection carried out for one–dimensional distributions of the cost and the effectiveness presents some inconveniences, suggesting that variable selection should then be carried out for the joint distribution of them.

6.2 The data and the Bayesian model

The data collected from n_i patients under treatment T_i, $i = 1, \ldots, k$, are given by the triple $(\mathbf{e}_i, \mathbf{c}_i, \mathbf{X})$, where $\mathbf{e}_i = (e_{i1}, \ldots, e_{in_i})^\top$ is a sample of the effectiveness, $\mathbf{c}_i = (c_{i1}, \ldots, c_{in_i})^\top$ a sample of the cost, and

$$\mathbf{X} = \begin{pmatrix} 1 & x_{11} & \cdots & x_{p1} \\ \vdots & \vdots & \ddots & \vdots \\ 1 & x_{1n_i} & \cdots & x_{pn_i} \end{pmatrix}$$

the design matrix, a matrix of dimensions $n_i \times (p+1)$, where the column $s+1$ contains the values $(x_{s1}, \ldots, x_{sn_i})^\top$ of the covariate x_s from the n_i patients under T_i, $s = 1, \ldots, p$. The number p denotes the potential number of covariates. Thus, the cost of the n_i patients are not identically distributed, and the same happens for the effectiveness. For this type of data, linear regression models are typically used (Nixon and Thompson, 2005; Hoch et al., 2006; Chib and Jacobi, 2007; Manca et al., 2007; Moreno et al., 2012, 2013b).

When the cost and the effectiveness are both continuous random variables we consider the following regression models: a first one in which c and e are independent and their distributions are normal; a second one in which c, conditional on e, is normal distributed, and the distribution e is also normal; and a third regression model in which the distribution of c, conditional on e, follows a lognormal distribution, and e is normal distributed. When the effectiveness e is a discrete random variable the normal model is replaced by the probit model.

Suitable priors for models and model parameters for the Bayesian variable selection procedure for the mentioned models will be presented in Section 6.3. For these priors, the procedure enjoys a very good sampling behavior for moderate and large sample sizes.

6.2.1 The independent normal–normal model

The model assumes that the cost and effectiveness of treatments T_i for $i = 1, \ldots, k$, are continuous independent random variables such that

$$c_i = \alpha_{i0} + \sum_{s=1}^{p} \alpha_{is} x_s + \varepsilon_i, \tag{6.1}$$

and

$$e_i = \beta_{i0} + \sum_{s=1}^{p} \beta_{is}x_s + \varepsilon_i', \tag{6.2}$$

where the random error ε_i and ε_i' follow the normal distributions $\mathcal{N}(\varepsilon_i|0,\sigma_i^2)$ and $\mathcal{N}(\varepsilon_i'|0,\tau_i^2)$, σ_i^2 and τ_i^2 are unknown variance errors, and $\boldsymbol{\alpha}_i = (\alpha_{i0},\ldots,\alpha_{ip})^\top$ and $\boldsymbol{\beta}_i = (\beta_{i0},\ldots,\beta_{ip})^\top$ are unknown regressor coefficients.

Thus, the model for the sample cost $\mathbf{c}_i = (c_{i1},\ldots,c_{in_i})^\top$ and effectiveness $\mathbf{e}_i = (e_{i1},\ldots,e_{in_i})^\top$ can be written as

$$\mathcal{N}_{n_i}(\mathbf{c}_i|\mathbf{X}\boldsymbol{\alpha}_i,\sigma_i^2\mathbf{I}_{n_i})\,\mathcal{N}_{n_i}(\mathbf{e}_i|\mathbf{X}\boldsymbol{\beta}_i,\tau_i^2\mathbf{I}_{n_i}).$$

To complete the specification of the Bayesian model for treatment T_i we need a prior distribution for the parameters $\boldsymbol{\alpha}_i, \sigma_i$ and $\boldsymbol{\beta}_i, \tau_i$. In the absence of subjective prior information on these parameters, we can use the reference priors

$$\pi^N(\boldsymbol{\alpha}_i,\sigma_i) \propto \frac{1}{\sigma_i}\mathbf{1}_{\mathbb{R}^{p+1}\times\mathbb{R}+}(\boldsymbol{\alpha}_i,\sigma_i) \text{ and } \pi^N(\boldsymbol{\beta}_i,\tau_i) \propto \frac{1}{\tau_i}\mathbf{1}_{\mathbb{R}^{p+1}\times\mathbb{R}+}(\boldsymbol{\beta}_i,\tau_i),$$

although, as we mentioned in Chapter 2, they are improper priors, and while this is not an inconvenience for estimating the parameters, they cannot be used for variable selection.

6.2.2 The dependent normal–normal model

The model assumes that the cost and the effectiveness of treatment T_i, $i = 1,\ldots,k$, are continuous random variables that can be written as

$$c_i = \alpha_{i0} + \sum_{s=1}^{p} \alpha_{is}x_s + \alpha_{i,p+1}\,e_i + \varepsilon_i, \tag{6.3}$$

and

$$e_i = \beta_{i0} + \sum_{s=1}^{p} \beta_{is}x_s + \varepsilon_i', \tag{6.4}$$

where the random error ε_i and ε_i' follow the normal distributions $\mathcal{N}(\varepsilon_i|0,\sigma_i^2)$ and $\mathcal{N}(\varepsilon_i'|0,\tau_i^2)$, σ_i^2 and τ_i^2 are unknown variance errors, and $\boldsymbol{\alpha}_i = (\alpha_{i0}, \alpha_{i1}, \ldots, \alpha_{ip}, \alpha_{i,p+1})^\top$ and $\boldsymbol{\beta}_i = (\beta_{i0}, \ldots, \beta_{ip})^\top$ are unknown regression coefficients.

We note that the cost model includes the effectiveness as a regressor. This simple form of conditioning the cost on the effectiveness has been considered by Willke et al. (1998), Vázquez–Polo et al. (2005b), Manca et al. (2007), Willan and Kowgier (2008) and Moreno et al. (2012, 2013b).

Therefore, the model for the sample cost $\mathbf{c}_i = (c_{i1}, \ldots, c_{in_i})^\top$ and effectiveness $\mathbf{e}_i = (e_{i1}, \ldots, e_{in_i})^\top$ can be written as

$$\mathcal{N}_{n_i}(\mathbf{c}_i | \mathbf{X}_c \boldsymbol{\alpha}_i, \sigma_i^2 \mathbf{I}_{n_i}) \times \mathcal{N}_{n_i}(\mathbf{e}_i | \mathbf{X}_e \boldsymbol{\beta}_i, \tau_i^2 \mathbf{I}_{n_i}), \tag{6.5}$$

where the design matrix $\mathbf{X}_c$ has dimensions $n_i \times (p+2)$ and is defined by adding to the design matrix $\mathbf{X}$ a column with the effectiveness data $\mathbf{e}_i$, that is,

$$\mathbf{X}_c = \begin{pmatrix} 1 & x_{11} & \cdots & x_{p1} & e_{i1} \\ \vdots & \vdots & \ddots & \vdots & \vdots \\ 1 & x_{1n_i} & \cdots & x_{pn_i} & e_{in_i} \end{pmatrix},$$

$\mathbf{X}_e = \mathbf{X}$, and $\mathbf{I}_{n_i}$ is the identity matrix of dimensions $n_i \times n_i$.

To complete the specification of the Bayesian model we again consider the reference priors

$$\pi^N(\boldsymbol{\alpha}_i, \sigma_i) \propto \frac{1}{\sigma_i} \mathbf{1}_{\mathbb{R}^{p+2} \times \mathbb{R}^+}(\boldsymbol{\alpha}_i, \sigma_i) \text{ and } \pi^N(\boldsymbol{\beta}_i, \tau_i) \propto \frac{1}{\tau_i} \mathbf{1}_{\mathbb{R}^{p+1} \times \mathbb{R}^+}(\boldsymbol{\beta}_i, \tau_i).$$

6.2.3   The dependent lognormal–normal model

Most of the time, the distribution of the cost exhibits a certain degree of skewness, so an asymmetric lognormal distribution is typically proposed as an appropriate distribution for the cost. Thus, it is typically assumed that the logarithm of the cost and the effectiveness of treatment T_i, $i = 1, \ldots, k$, can be written as

$$\log(c_i) = \gamma_{i0} + \sum_{s=1}^{p} \gamma_{is} x_s + \gamma_{i,p+1} e_i + \varepsilon_i, \quad \varepsilon_i \sim \mathcal{N}(\varepsilon_i | 0, \eta_i^2), \tag{6.6}$$

$$e_i = \beta_{i0} + \sum_{s=1}^{p} \beta_{is} x_s + \varepsilon_i', \qquad \varepsilon_i' \sim \mathcal{N}(\varepsilon_i'|0, \tau_i^2), \qquad (6.7)$$

where $\boldsymbol{\gamma}_i = (\gamma_{i0}, \ldots, \gamma_{ip}, \gamma_{i,p+1})^\top$ are unknown regression coefficients and η_i^2 the unknown variance error of the logarithm of the cost. This formulation (6.6) and (6.7) means that the cost is partially explained by a set of covariates that includes the effectiveness, which has a multiplicative effect on it.

Therefore, the model for the sample cost $\mathbf{c}_i = (c_{i1}, \ldots, c_{in_i})^\top$ and effectiveness $\mathbf{e}_i = (e_{i1}, \ldots, e_{in_i})^\top$ is now written as

$$\Lambda_{n_i}(\mathbf{c}_i | \mathbf{X}_c \boldsymbol{\gamma}_i, \eta_i^2 \mathbf{I}_{n_i}) \times \mathcal{N}_{n_i}(\mathbf{e}_i | \mathbf{X}_e \boldsymbol{\beta}_i, \tau_i^2 \mathbf{I}_{n_i}), \qquad (6.8)$$

where Λ_{n_i} denotes the multivariate lognormal distribution.

The priors for $\boldsymbol{\gamma}_i, \eta_i$ and $\boldsymbol{\beta}_i, \tau_i$ are the reference priors given in Section 6.2.1.

6.2.4 The probit sampling model

When the effectiveness e of treatment T is a discrete $0-1$ random variable indicating nonsuccess and success, the following regression model in the presence of the covariates $\mathbf{x}^\top = (1, x_1, \ldots, x_p)$ of the patients can be assumed (León–Novelo et al., 2012). The variable e follows the Bernoulli distribution

$$\mathrm{Be}(e|\theta) = \theta^e (1-\theta)^{1-e}, \quad e = 0, 1,$$

and the probability of success θ is such that

$$\theta = \Phi(-\mathbf{x}^\top \boldsymbol{\beta}),$$

where

$$\Phi(-\mathbf{x}^\top \boldsymbol{\beta}) = \frac{1}{\sqrt{2\pi}} \int_{-\infty}^{-\mathbf{x}^\top \boldsymbol{\beta}} \exp\left\{-\frac{u^2}{2}\right\} du,$$

and $\boldsymbol{\beta}^\top = (\beta_0, \beta_1, \ldots, \beta_p)$ is an unknown vector of regression parameters.

That is, we are assuming that θ is a function of the covariates and it is given by the cumulative standard normal distribution at point $-\mathbf{x}^\top\boldsymbol{\beta} = -(\beta_0 + \sum_{j=1}^p x_j\beta_j)$.

This probit model for e can be thought of as a normal regression model with incomplete sampling information. Let y be a latent random variable with the normal distribution $\mathcal{N}(y|\mathbf{x}^\top\boldsymbol{\beta}, 1)$ for which only its sign can be observed, and define e as

$$e = \begin{cases} 0, & \text{if } y > 0, \\ 1, & \text{if } y \leq 0. \end{cases}$$

This way, a sample $\mathbf{y} = (y_1, \ldots, y_n)$ of size n of the latent variable y is identified with the sample $\mathbf{e} = (e_1, \ldots, e_n)^\top$ of effectiveness from n patients under treatment T by the equality

$$(\text{sign}(y_1), \ldots, \text{sign}(y_n))^\top = (e_1, \ldots, e_n)^\top.$$

Further, we have that

$$\theta = \Pr(y \leq 0) = \int_{-\infty}^0 \mathcal{N}(y|\mathbf{x}^\top\boldsymbol{\beta}, 1)\, dy = \frac{1}{\sqrt{2\pi}} \int_{-\infty}^{-\mathbf{x}^\top\boldsymbol{\beta}} \exp\left\{-\frac{u^2}{2}\right\}\, du.$$

The idea behind this artificial construction is that of using the well–established Bayesian variable selection procedure for the normal model for variable selection in probit models. This point will be considered in Section 6.3.7.

6.3 Bayesian variable selection

For the construction of the patient subgroups of a generic treatment T, the covariates we include in the regression model of the cost c and the effectiveness e are crucial. The question is whether p, the initial number of covariates, can be reduced by selecting a subset of them that has an influence on (c, e) based on the sampling information $\mathbf{c} = (c_1, \ldots, c_n)^\top$ and $\mathbf{e} = (e_1, \ldots, e_n)^\top$ from patients under treatment T.

Several frequentist and Bayesian statistical solutions to this important variable selection problem have been proposed in its vast literature.

This is in nature a decision problem with sampling information $(\mathbf{c}, \mathbf{e})$, where the decision space is the set of 2^p regression models $\mathfrak{M}$ for $\mathbf{c}$ and $\mathbf{e}$ defined by all possible subsets of the original set of p covariates. This set can be written as the union

$$\mathfrak{M} = \bigcup_{k=0}^{p} \mathfrak{M}_k$$

where $\mathfrak{M}_k$ is the set of regression models with k regressors. The reward of model M_k in $\mathfrak{M}$ is their posterior probability conditional on the sample $\mathbf{y} = (\mathbf{c}, \mathbf{e})$. The utility function we choose is a $0 - 1$ function whose meaning is to win 1 unit when making the decision of choosing the true model for $\mathbf{y}$, and 0 otherwise.

It is immediate to see that the optimal model, the model having the maximum expected utility, is the one having the highest posterior probability. Thus, the quantity of interest in variable selection is the posterior distribution of the models in $\mathfrak{M}$, conditional on the sample $\mathbf{y} = (\mathbf{c}, \mathbf{e})$ and the design matrices $\{\mathbf{X}_k, \ k = 1, \ldots, 2^p\}$.

Here we give a Bayesian procedure for variable selection in normal regression models for some prior distributions for models and model parameters commonly used. Popular prior distributions for the model parameters are Zellner's $g-$priors, mixtures of Zellner's $g-$priors, and the intrinsic priors. These priors are presented in Section 6.3.4. Further, a firm candidate prior for models is the hierarchical uniform prior, which is presented in Section 6.3.5.

It can be shown that the posterior model probability for the hierarchical uniform prior for the models and either the Zellner or the intrinsic priors for the model parameters exhibits very good sampling behavior. Our favorite priors for model parameters are the intrinsic priors because they are completely automatic, no assumption on common parameters for the models is required, and the computation of the posterior model probability is simpler for the intrinsic priors than for Zellner's priors.

Since the quantity of interest in variable selection is the posterior model probability in a space of 2^p regression models, one might worry when observing that the number of models grows exponentially with p. A number of potential regressors p that are not very large, yields a certainly large number of models. For instance, for $p = 20$ the number of models is as large as 1048576. This exponential growing provokes a serious numerical problem even for moderately large values of p. In Casella and Moreno (2006) a Metropolis–Hasting stochastic search algorithm was used to overcome the numerical problem for p moderately large. The idea behind the search is that of reducing the computation of posterior model probabilities to a subset of models. This subset is defined by a sequence of models such that their posterior model probability grows sequentially as the Metropolis–Hasting steps go forward. Unfortunately, there is no guarantee that we get the optimal model, but certainly we typically get a model with high enough posterior probability.

In this section we give the Bayesian variable selection procedure for a generic normal class of linear models, and the results will be applied to the normal–normal models, the lognormal–normal models, and the probit regression models.

6.3.1   Notation

Let $(\mathbf{y}, \mathbf{X})$ be the observed responses and covariates from n independent patients. We assume that $(\mathbf{y}, \mathbf{X})$ follows the n dimensional normal distribution $\mathcal{N}_n(\mathbf{y}|\mathbf{X}\boldsymbol{\beta}_{p+1}, \sigma_p^2\mathbf{I}_n)$. This model, which includes all covariates, is called the full model, and is denoted as M_p. The model $\mathcal{N}_n(\mathbf{y}|\beta_0\mathbf{1}_n, \sigma_0^2\mathbf{I}_n)$, which includes no covariates, is called *the intercept–only model*, and is denoted as M_0.

By M_j we denote a generic model $\mathcal{N}_n(\mathbf{y}|\mathbf{X}_{j+1}\boldsymbol{\beta}_{j+1}, \sigma_j^2\mathbf{I}_n)$ with j covariates, $0 \le j \le p$, where $\boldsymbol{\beta}_{j+1} = (\beta_0, \beta_1, \ldots, \beta_j)^\mathsf{T}$, $\mathbf{X}_{j+1}$ is an $n \times (j+1)$ submatrix of $\mathbf{X}$ formed with j specific covariates, and σ_j^2 is the variance error. The number of models M_j with j regressors is $p!/(j!(p-j)!)$, and the set of them is denoted as $\mathfrak{M}_j$. The class of all possible regression models with at most p regressors is $\mathfrak{M} = \cup_{j=0}^{p}\mathfrak{M}_j$. We remark that the regression coefficients change across models, although for simplicity we use the same alphabetical notation.

To complete the sampling models we need a prior for models in $\mathfrak{M}$ and for model parameters. Thus, for a generic model M_j we need a prior $\pi(\boldsymbol{\beta}_{j+1}, \sigma_j, M_j)$. It is convenient to decompose this prior as

$$\pi(\boldsymbol{\beta}_{j+1}, \sigma_j, M_j) = \pi(\boldsymbol{\beta}_{j+1}, \sigma_j | M_j)\pi(M_j)$$

for $M_j \in \mathfrak{M}_j$, $(\boldsymbol{\beta}_{j+1}, \sigma_j) \in \mathbb{R}^{j+1} \times \mathbb{R}^+$.

6.3.2 Posterior model probability

Given the sample $(\mathbf{y}, \mathbf{X})$ from a model in $\mathfrak{M}$ it follows from the Bayes theorem that the posterior probability in the class $\mathfrak{M}$ of model M_j, which contains j specific covariates, is given by

$$\Pr(M_j | \mathbf{y}, \mathbf{X}) = \frac{m_j(\mathbf{y}, \mathbf{X})\,\pi(M_j)}{\sum_{i=0}^{p}\sum_{M_i \in \mathfrak{M}_i} m_i(\mathbf{y}, \mathbf{X})\,\pi(M_i)}, \qquad (6.9)$$

where

$$m_i(\mathbf{y}, \mathbf{X}) = \int_{\mathbb{R}^{i+1}} \int_0^{\infty} \mathcal{N}_n(\mathbf{y} | \mathbf{X}_{i+1}\boldsymbol{\beta}_{i+1}, \sigma_i^2 \mathbf{I}_n)\pi(\boldsymbol{\beta}_{i+1}, \sigma_i | M_i)\,d\boldsymbol{\beta}_{i+1}\,d\sigma_i$$

denotes the marginal distribution of the sample under model M_i.

If we divide the numerator and the denominator of (6.9) by the marginal of the data under the intercept–only model $m_0(\mathbf{y}, \mathbf{X})$, the posterior probability of model M_j becomes

$$\Pr(M_j | \mathbf{y}, \mathbf{X}) = \frac{B_{j0}(\mathbf{y}, \mathbf{X})\pi(M_j)}{\sum_{i=0}^{p}\sum_{M_i \in \mathfrak{M}_i} B_{i0}(\mathbf{y}, \mathbf{X})\,\pi(M_i)}, \qquad (6.10)$$

where

$$B_{i0}(\mathbf{y}, \mathbf{X}) = \frac{m_i(\mathbf{y}, \mathbf{X})}{m_0(\mathbf{y}, \mathbf{X})}$$

is the Bayes factor for comparing models M_i and M_0. We note that the Bayes factor $B_{i0}(\mathbf{y}, \mathbf{X})$ is simply the ratio of the likelihood of model M_i and model M_0 for the data $(\mathbf{y}, \mathbf{X})$. We also note that model M_0 is nested in model M_i.

The advantage of writing the posterior model probability as we do in (6.10) is that the Bayes factors only involve nested models, and it is known that the Bayes factor for nested models enjoys, under mild conditions, excellent asymptotic properties.

On the other hand, Bayes factors do not depend on the prior for models but only on the prior for model parameters. We recall that improper priors for model parameters leave the Bayes factor defined up to an arbitrary multiplicative constant, and thus they cannot be used for computing Bayes factors. Priors for model parameters suitable for computing Bayes factors for variable selection in normal regression models are given in the next two sections. We also give the prior for the models by which to compute the posterior probability of the models.

6.3.3   The hierarchical uniform prior for models

Since $\mathfrak{M}$ is a discrete space with 2^p models, a natural default prior over this space is the uniform prior, but, as Moreno et al. (2015) showed, it is not necessarily a good prior. A generalization of the uniform prior is the parametric class of Bernoulli priors (George and McCulloch, 1993, 1997). For this class of priors the probability of a generic model M_j containing j out of p regressors, $j \leq p$, is given by

$$\pi(M_j|\theta) = \theta^j(1 - \theta)^{p-j}, \quad 0 \leq \theta \leq 1,$$

where θ is an unknown hyperparameter whose meaning is the probability of inclusion of a regressor in the model. We note that the model prior $\pi(M_j|\theta)$ assigns the same probability to models with the same dimension, that is, it is uniform on $\mathfrak{M}_j$. Further, for $\theta = 1/2$, the uniform prior $\pi(M_j|1/2) = 2^{-p}$ on $\mathfrak{M}$ is obtained.

If we further assume a uniform distribution for θ, the unconditional probability $\pi^{\text{HU}}(M_j)$ of model M_j is given by

$$\pi^{\text{HU}}(M_j) = \int_0^1 \theta^j(1 - \theta)^{p-j}d\theta = \binom{p}{j}^{-1}\frac{1}{p+1}. \qquad (6.11)$$

If we decompose this probability as

$$\pi^{\text{HU}}(M_j) = \pi^{\text{HU}}(M_j|\mathfrak{M}_j)\pi^{\text{HU}}(\mathfrak{M}_j),$$

it follows that the model prior distribution, conditional on the class $\mathfrak{M}_j$, is uniform, and the marginal over the classes $\{\mathfrak{M}_j, \ j = 0, 1, \ldots, p\}$ is also uniform. Then, it seems appropriate to call to this prior the hierarchical uniform prior (Moreno et al., 2015). It can be shown that for variable selection, the hierarchical prior $\pi^{\mathrm{HU}}(M_j)$ outperforms the behavior of the prior $\pi(M_j|\theta)$ for any value of θ.

6.3.4   Zellner's g−priors for model parameters

Zellner's g−priors were introduced by Zellner and Siow (1980) and Zellner (1986). A simplification of the regression models when using the g−priors for model selection is obtained by assuming that the intercept and the variance error are common parameters to all models. This reduces the number of parameters involved in the comparison of model M_j *versus* model M_0 from $j+4$ to $j+2$. According to this restriction the regression parameters of a generic model M_j will be denoted as $(\beta_0, \boldsymbol{\beta}_j)^{\mathsf{T}} = (\beta_0, \beta_1, \ldots, \beta_j)^{\mathsf{T}}$ and the variance error as σ^2, where β_0 and σ are common to all models.

For a sample $(\mathbf{y}, \mathbf{X})$, most references to g−priors in the variable selection literature (Berger and Pericchi, 2001; Clyde and George, 2004; Clyde et al., 1998; George and Foster, 2000; Fernández et al., 2001; Hansen and Yu, 2001; Liang et al., 2008, among others), refer to them as the pair $\pi^N(\beta_0, \sigma)$ and $\pi(\boldsymbol{\beta}_j|\sigma, g)$, where

$$\pi^N(\beta_0, \sigma) = \frac{k}{\sigma} \mathbf{1}_{\mathbb{R} \times \mathbb{R}^+}(\beta_0, \sigma)$$

is the reference prior and k an arbitrary positive constant, and

$$\pi(\boldsymbol{\beta}_j|\sigma, g) = \mathcal{N}_j(\boldsymbol{\beta}_j|\mathbf{0}_j, g\sigma^2(\mathbf{X}_j^{\mathsf{T}}\mathbf{X}_j)^{-1}).$$

In this expression, $\mathbf{0}_j$ denotes the column vector of zeros of dimension j, $g > 0$ is an unknown positive hyperparameter, and $\mathbf{X}_j$ the matrix of dimensions $n \times j$ resulting from suppressing the first column in the design matrix $\mathbf{X}_{j+1}$ of the original formulation of the regression model M_j.

The conjugate property of these priors makes the expression of the Bayes factor quite simple, and it is well known that the hyperparameter g plays an important role in the behavior of the Bayes factor. Several values for g have been suggested, although none of them satisfies all the reasonable requirements (Berger and Pericchi, 2001; Clyde and George, 2004; Clyde et al., 1998; George and Foster, 2000; Fernández et al., 2001; Hansen and Yu, 2001; Liang et al., 2008). For instance, large g values induce the Lindley–Bartlett paradox (Bartlett, 1957), and a fixed value for g induces inconsistency, which can be corrected if g was dependent on n.

We consider two versions of the g−prior. The first is the one obtained for $g = n$, which is justified on the grounds that it provides a consistent Bayes factor, and it is a *unit information prior* (Kass and Wasserman, 1996). The second g−prior version was derived for avoiding an incoherent property of the g−prior detected by Berger and Pericchi (2001). To avoid this incoherent behavior it was suggested to integrate out g from the conditional g−priors $\{ \pi(\boldsymbol{\beta}_j|\sigma, g),\ g > 0 \}$ to obtain the mixture of g−priors

$$\pi^{\mathrm{mix}}(\boldsymbol{\beta}_j|\sigma) = \int_0^\infty \pi(\boldsymbol{\beta}_j|\sigma, g)\, \pi(g)\, dg,$$

where $\pi(g)$ is the inverse–gamma density

$$\pi(g) = \frac{(n/2)^{1/2}}{\Gamma(1/2)} g^{-3/2} \exp\left(-\frac{n}{2g}\right).$$

This mixture has been considered by some other authors, including Clyde and George (2004), Liang et al. (2008), Scott and Berger (2010) and Moreno et al. (2015).

6.3.5   Intrinsic priors for model parameters

Intrinsic priors for computing Bayes factors in variable selection have been used by Moreno and Girón (2005), Casella and Moreno (2006), Girón et al. (2006), León–Novelo et al. (2012), Consonni et al. (2013), Moreno et al. (2015), among others.

The standard intrinsic method for comparing the null model M_0 *versus* the alternative M_j starts with the improper reference priors for the parameters of these models (α_0, σ_0) and $(\boldsymbol{\beta}_{j+1}, \sigma_j)$, that is, $\pi^N(\alpha_0, \sigma_0) = c_0/\sigma_0$ and $\pi^N(\boldsymbol{\beta}_{j+1}, \sigma_j) = c_j/\sigma_j$, and provides a proper intrinsic prior for the parameters $(\boldsymbol{\beta}_{j+1}, \sigma_j)$ conditional on the parameter (α_0, σ_0) of the model M_0, as

$$\pi^I(\boldsymbol{\beta}_{j+1}, \sigma_j | \alpha_0, \sigma_0) = \mathcal{N}_{j+1}(\boldsymbol{\beta}_{j+1} | \tilde{\boldsymbol{\alpha}}_0, (\sigma_j^2 + \sigma_0^2) \mathbf{W}_{j+1}^{-1}) \, HC^+(\sigma_j | \sigma_0),$$

where $\tilde{\boldsymbol{\alpha}}_0 = (\alpha_0, \mathbf{0}_j^\top)^\top$, $\mathbf{W}_{j+1}^{-1} = \dfrac{n}{j+2}(\mathbf{X}_{j+1}^\top \mathbf{X}_{j+1})^{-1}$, and

$$HC^+(\sigma_j | \sigma_0) = \frac{2}{\pi} \frac{\sigma_0}{\sigma_j^2 + \sigma_0^2}$$

is the half Cauchy distribution on $\mathbb{R}^+$ with location parameter 0 and scale σ_0. The unconditional intrinsic prior for $(\boldsymbol{\beta}_{j+1}, \sigma_j)$ is then given by

$$\pi^I(\boldsymbol{\beta}_{j+1}, \sigma_j) = \int_{-\infty}^{\infty} \int_0^{\infty} \pi^I(\boldsymbol{\beta}_{j+1}, \sigma_j | \alpha_0, \sigma_0) \, \pi^N(\alpha_0, \sigma_0) \, d\alpha_0 \, d\sigma_0.$$

For comparing model M_j *versus* M_0 the intrinsic priors are the pair $(\pi^I(\boldsymbol{\beta}_{j+1}, \sigma_j), \pi^N(\alpha_0, \sigma_0))$. We remark that $\pi^I(\boldsymbol{\beta}_{j+1}, \sigma_j)$ depends on the arbitrary constant c_0 that cancels out in the Bayes factor $B_{j0}(\mathbf{y}, \mathbf{X})$, and hence no tuning hyperparameters have to be adjusted.

Thus, the intrinsic priors are automatically constructed from the sampling models and the reference priors, and when using the intrinsic prior we do not need to assume that the intercept and the variance error are common parameters to all models.

A summary of the properties of the Bayes factors for the two versions of Zellner's g−priors and the intrinsic priors for variable selection in a normal regression are given in Moreno et al. (2015).

6.3.6   Bayes factors for normal linear models

For the data $(\mathbf{y}, \mathbf{X})$, it can easily be seen that the Bayes factor for comparing M_j *versus* M_0 for the g−prior with $g = n$ is given by

$$B_{j0}^{g=n}(\mathbf{y}, \mathbf{X}) = \frac{(1+n)^{(n-j-1)/2}}{(1+n\,\mathcal{B}_{j0})^{(n-1)/2}},$$ (6.12)

for the mixture of $g-$priors by

$$B_{j0}^{\text{mix}}(\mathbf{y}, \mathbf{X}) = \frac{(n/2)^{1/2}}{\Gamma(1/2)} \int_0^\infty \frac{(1+g)^{(n-j-1)/2}}{(1+g\,\mathcal{B}_{j0})^{(n-1)/2}}\, g^{-3/2} \exp\left(-\frac{n}{2g}\right)\, dg,$$ (6.13)

and for the intrinsic priors by

$$B_{j0}^{\text{I}}(\mathbf{y}, \mathbf{X}) = \frac{2}{\pi}(j+2)^{j/2} \int_0^{\pi/2} \frac{\sin^j \varphi\, (n + (j+2)\sin^2 \varphi)^{(n-j-1)/2}}{(n\,\mathcal{B}_{j0} + (j+2)\sin^2 \varphi)^{(n-1)/2}}\, d\varphi.$$ (6.14)

The integrals on $(0, \infty)$ in (6.13) and on $(0, \pi/2)$ in (6.14) do not have explicit expressions but need numerical integration.

We note that all these Bayes factors depend on the data through the statistic $\mathcal{B}_{j0}$, which is the ratio of the square sum of the residuals of models M_j and M_0, that is

$$\mathcal{B}_{j0} = \frac{\mathbf{y}^\top (\mathbf{I} - \mathbf{H}_j)\mathbf{y}}{\mathbf{y}^\top \left(\mathbf{I} - \frac{1}{n}\mathbf{1}_n\mathbf{1}_n^\top\right)\mathbf{y}},$$

where $\mathbf{H}_j = \mathbf{X}_j(\mathbf{X}_j^\top \mathbf{X}_j)^{-1}\mathbf{X}_j^\top$ is the hat matrix associated with $\mathbf{X}_j$.

In the next example we use simulated data to illustrate the performance of the three Bayesian variable selection procedures with Bayes factors (6.12), (6.13) and (6.14), and the hierarchical uniform model prior given in (6.11).

Example 6.1. *Let $\mathcal{N}(y|\mathbf{x}\boldsymbol{\beta}, 1)$ be the full normal distribution for the random variable y, where $\mathbf{x} = (1, x_1, \ldots, x_6)$ is the vector of six deterministic covariates and $\boldsymbol{\beta} = (\beta_0, \beta_1, \ldots, \beta_6)^\top$ the regression coefficients. We consider the class of $2^6 = 64$ normal models defined by the subsets of the original six covariates.*

We simulate a vector of covariates $\mathbf{x}_t$ from a uniform distribution on $(0, 10)$ and set $\boldsymbol{\beta}_t = (1, 0, 1, 0, 1, 0, 1)$. Then, a value of the variable y is simulated from the normal distribution $\mathcal{N}(y|\mathbf{x}_t\boldsymbol{\beta}_t, 1)$. We note that *only covariates* x_2, x_4, *and* x_6 *enter in the true model that generated* y. *We repeat the simulation of the covariates and the response variable 30 times, and the simulated data are written as* $(\mathbf{y}, \mathbf{X}_6)$ *where* $\mathbf{y} = (y_1, \ldots, y_{30})^\top$ *and* $\mathbf{X}_6$ *is a matrix of dimension* 30×6 *containing the regressors.*

With these data we compute the posterior probabilities of the 64 models for the following priors: the hierarchical uniform prior $\pi^{HU}(M)$ for models $M \in \mathfrak{M}$, Zellner's g−prior for $g = 30$, the mixture of g−priors, and the intrinsic priors for the model parameters.

We repeat the simulations and computations 100 times, and the mean across simulations of the posterior probabilities of the true model for the priors for the model parameters are displayed in the second column in Table 6.1.

TABLE 6.1

Priors (first column) and posterior probabilities of the true model (second column).

Prior for model parameters	Mean of the posterior probabilities of the true model across simulations
g−prior with $g = 30$	0.34
Mixture of g−priors	0.55
Intrinsic priors	0.57

From Table 6.1 it follows that the largest mean across simulations of the posterior probability of the true model is obtained for the intrinsic priors, and the second largest for the mixture of g−priors. We recall that these posterior probabilities are computed in the space of 64 models, and hence their values are really large.

We also found that the proportion of times the true model has the highest posterior probability across simulations was 91% for the g−prior and 92% for either the mixture of g−priors or intrinsic priors.

We remark that Table 6.1 has been constructed for samples of moderate size, $n = 30$. The posterior probability of the true model certainly changes as n increases, and for large enough n, it will be close to one for any of the priors used. This result follows, under mild conditions, from the posterior model consistency of the Bayesian variable selection procedure for the above priors in Moreno et al. (2015).

6.3.7 Bayes factors for probit models

The Bayes factor for the probit model and the intrinsic priors for the regression parameters are computed from the Bayes factors for normal linear models as follows. We consider the latent random variable y, for which we are only able to observe its sign, whose distribution is $\mathcal{N}(y|\mathbf{x}^\top\boldsymbol{\beta}, 1)$. Let $\mathcal{N}(y|\alpha, 1)$ denote the intercept–only model.

For a the latent sample $\mathbf{y} = (y_1, \ldots, y_{n_i})$, the marginal $m_j(\mathbf{y}, \mathbf{X}_j)$ of the data for the normal model with j regressors and the intrinsic prior $\pi^I(\boldsymbol{\beta}_j|\alpha)\pi^N(\alpha)$ is given by

$$
m_j(\mathbf{y}, \mathbf{X}_j) = \int_{\mathbb{R}^{j+1}} \int_{\mathbb{R}} \mathcal{N}(\mathbf{y}|\mathbf{X}_j\boldsymbol{\beta}_j, \mathbf{I}_n)\pi^I(\boldsymbol{\beta}_j|\alpha)\pi^N(\alpha)\, d\alpha\, d\boldsymbol{\beta}_j
$$
$$
= c \int_{\mathbb{R}^{j+1}} \int_{\mathbb{R}} \mathcal{N}(\mathbf{y}|\mathbf{X}_j\boldsymbol{\beta}_j, \mathbf{I}_n)\mathcal{N}_{j+1}(\boldsymbol{\beta}_{j+1}|\tilde{\alpha}_0, (\sigma_j^2 + \sigma_0^2)\mathbf{W}_{j+1}^{-1})
$$
$$
\times HC^+(\sigma_j|\sigma_0)\, d\alpha\, d\boldsymbol{\beta}_j
$$

where c is a positive arbitrary constant, $\mathbf{X}_j$ a matrix of dimension $n_i \times (j+1)$, and $\boldsymbol{\beta}_j$ a column vector of dimension $j+1$. If the conditional intrinsic prior $\pi^I(\boldsymbol{\beta}_j|\alpha)$ is replaced with any version of the g–prior, a different but close marginal is obtained.

The marginal under the intercept–only model for $\pi^N(\alpha)$ is

$$
m_0(\mathbf{y}) = \int_{\mathbb{R}} \mathcal{N}(\mathbf{y}|\alpha\mathbf{1}_n, \mathbf{I}_n)\pi(\alpha)\, d\alpha = c \int_{\mathbb{R}} \mathcal{N}(\mathbf{y}|\alpha\mathbf{1}_n, \mathbf{I}_n)\, d\alpha,
$$

where $\mathbf{1}_n = (1, \ldots, 1)^\top$.

Let $\mathbf{e}_i = (e_{i1}, \ldots, e_{in_i})$ be a sample of effectiveness of the treatment T_i. Since $\mathbf{e}_i = (\text{sign}(y_1), \ldots, \text{sign}(y_{n_i}))$, the marginal of $(\mathbf{e}_i, \mathbf{X}_j)$ under model M_j is given by

$$m_j(\mathbf{e}_i, \mathbf{X}_j) = \int_{A_1 \times \dots \times A_{n_i}} m_j(\mathbf{y}, \mathbf{X}_j) \, d\mathbf{y}$$

where

$$A_i = \begin{cases} (0, \infty), & \text{if } e_i = 0, \\ (-\infty, 0), & \text{if } e_i = 1. \end{cases}$$

Likewise, the marginal under the intercept–only model is given by

$$m_0(\mathbf{e}_i) = \int_{A_1 \times \dots \times A_{n_i}} m_0(\mathbf{y}, \mathbf{X}_j) \, d\mathbf{y}.$$

The integration on $A_1 \times \dots \times A_{n_i}$ does not have an explicit form, and it has to be computed numerically. This way we obtain the Bayes factor to compare model M_j against the intercept–only model M_0 as

$$B_{j0}(\mathbf{c}_j, \mathbf{X}) = \frac{m_j(\mathbf{e}_i, \mathbf{X}_j)}{m_0(\mathbf{e}_i)}.$$

The posterior probability of model M_j for $j = 1, \dots, 2^p$, is obtained from expression (6.10). This set of posterior model probabilities is the solution to the variable selection problem for the probit model. For more details on the subject, see León–Novelo et al. (2012).

Example 6.2. *The data in this example were obtained from a study to compare three methadone maintenance programs: high, medium and low intensity, for opioid–addicted patients (Puigdollers et al., 2003). In those programs, a 12–month follow–up study of 586 patients beginning methadone treatment at five drug care centers in Barcelona was performed. For illustrating the variable selection procedure in probit models we focus on the high–intensity program (treatment T).*

The effectiveness is a binary variable e that takes value 0 if the patient leaves the treatment, and 1 otherwise.

The potential set of covariates are the age of patient, the number of years of consumption, the Nottingham Health Profile (NHP) at the beginning of the treatment period, and the gender.

The NHP is a generic and standardized measure of health–related quality of life in terms of the subjective emotional, functional and social impact of disease. It ranges from 0 (normal health) to 100 (very poor health). We use the value of the NHP at the beginning of the treatment period. The gender takes a value 1 for female, and 0 otherwise. Table 6.2 gives a summary of these covariates.

TABLE 6.2

Summary of the effectiveness and covariates in Example 6.2 (mean and standard deviations, in parentheses).

Effectiveness ($e = 1$)	0.748
Sample size	$n_2 = 155$
Age (x_1)	30.53 (6.43)
Years (x_2)	10.17 (5.78)
NHP (x_3)	42.47 (23.55)
Gender (x_4)	0.28

Table 6.3 shows the top models and their probabilities after the application to the objective Bayesian probit model of the variable selection procedure.

TABLE 6.3

Top models in Example 6.2.

Model	Posterior Probability
{ Intercept, x_1, x_3 }	0.493
{ Intercept, x_1, x_3, x_4 }	0.254
{ Intercept, x_1, x_2, x_3 }	0.085
{ Intercept, x_1, x_2, x_3, x_4 }	0.051
the rest	< 0.05

From Table 6.3 it follows that not only the variables x_1 and x_3 define the top model but they are in the four top models. The posterior *probability of selecting x_1 is 0.99 and x_3, 0.88. The rest of the variables, x_2 and x_4, have marginal posterior probabilities 0.18 and 0.35, respectively.*

6.4   Bayesian predictive distribution of the net benefit

In this section we assume that the effectiveness e and cost c follow either the normal–normal distribution (6.5) or the lognormal–normal (6.8), and compute the Bayesian predictive distribution of the net benefit z, conditional on R, the data, and the covariate vector $\mathbf{x}$.

6.4.1   The normal–normal case

Suppose that from the application of the variable selection procedure to the original regression models for the sample of effectiveness and costs $\mathbf{e} = (e_1, \ldots, e_n)^\top$ and $\mathbf{c} = (c_1, \ldots, c_n)^\top$ of n patients, we end up with the models $\mathcal{N}_n(\mathbf{e}|\mathbf{X_e}\boldsymbol{\beta}, \tau^2 \mathbf{I}_n)$ and $\mathcal{N}_n(\mathbf{c}|\mathbf{X_c}\boldsymbol{\alpha}, \sigma^2 \mathbf{I}_n)$, where $\mathbf{X_e}$ and $\mathbf{X_c}$ are the design matrices of the selected regressors having dimensions $n \times m$ and $n \times m'$, respectively. In general, we will have that $m \neq m'$, $m \leq p$ and $m' \leq p + 1$.

If the effectiveness e is an influential regressor for the cost c, the matrix $\mathbf{X_c}$ will contain column $\mathbf{e}$.

The aim of this section is that of finding the Bayesian predictive distribution of the net benefit of a treatment for the dependent normal–normal model for the cost c and the effectiveness e. Assuming the reference prior densities for the regression parameters and the variance errors, that is,

$$\pi^N(\boldsymbol{\alpha}, \sigma) \propto \frac{1}{\sigma}, \qquad \pi^N(\boldsymbol{\beta}, \tau) \propto \frac{1}{\tau},$$

the posterior density of parameter $(\boldsymbol{\alpha}, \sigma)$, conditional on $(\mathbf{c}, \mathbf{X_c})$, turns out to be

$$\pi(\boldsymbol{\alpha}, \sigma | \mathbf{c}, \mathbf{X_c}) = \mathcal{N}_{m'} \text{InvGa}\left(\boldsymbol{\alpha}, \sigma | \widehat{\boldsymbol{\alpha}}, (\mathbf{X_c^\top X_c})^{-1}; \frac{\nu'}{2}, \frac{\nu'}{2} s_c^2\right), \qquad (6.15)$$

and the posterior distribution of $(\boldsymbol{\beta}, \tau^2)$, conditional on $(\mathbf{e}, \mathbf{X_e})$, is

$$\pi(\boldsymbol{\beta}, \tau | \mathbf{e}, \mathbf{X_e}) = \mathcal{N}_m \mathrm{InvGa}(\boldsymbol{\beta}, \tau^2 | \widehat{\boldsymbol{\beta}}, (\mathbf{X_e^\top X_e})^{-1}; \frac{\nu}{2}, \frac{\nu}{2} s_e^2), \qquad (6.16)$$

where $\mathcal{N}_q \mathrm{InvGa}$ represents a $q-$variate Normal–inverted–Gamma distribution, $\widehat{\boldsymbol{\alpha}} = (\mathbf{X_c^\top X_c})^{-1} \mathbf{X_c^\top c}$ and $\widehat{\boldsymbol{\beta}} = (\mathbf{X_e^\top X_e})^{-1} \mathbf{X_e^\top e}$ are the MLE estimators of $\boldsymbol{\alpha}$ and $\boldsymbol{\beta}$, respectively, $\nu' = n - m'$ and $\nu = n - m$ are degrees of freedom, $s_c^2 = RSS_c / \nu'$ and $s_e^2 = RSS_e / \nu$ are the usual unbiased estimators of the variances σ^2 and τ^2, respectively, and $RSS_c = (\mathbf{c} - \mathbf{X_c}\boldsymbol{\alpha})^\top (\mathbf{c} - \mathbf{X_c}\boldsymbol{\alpha})$ and $RSS_e = (\mathbf{e} - \mathbf{X_e}\boldsymbol{\beta})^\top (\mathbf{e} - \mathbf{X_e}\boldsymbol{\beta})$.

Before finding the Bayesian predictive distribution of z, conditional on the samples $(\mathbf{c}, \mathbf{e})$, we need the results in Lemmas 6.1 and 6.2.

Lemma 6.1. *Let* $\mathbf{y}_1$ *and* $\mathbf{y}_2$ *be random vectors of dimensions* n_1 *and* n_2, *and* τ *a nonnegative random variable with joint distribution*

$$\mathcal{N}_{n_1}(\mathbf{y}_1 | \mathbf{A}\mathbf{y}_2 + \mathbf{b}, \tau \mathbf{S}) \times \mathcal{N}_{n_2} InGa(\mathbf{y}_2, \tau | \mathbf{m}, \mathbf{V}; \xi, \upsilon).$$

Then, the joint distribution of $(\mathbf{y}_1, \tau)$ *is given by*

$$\mathcal{N}_{n_1} InGa(\mathbf{y}_1, \tau | \mathbf{A}\mathbf{m} + \mathbf{b}, \mathbf{S} + \mathbf{A}\mathbf{V}\mathbf{A}^\top; \xi, \upsilon).$$

Proof. The joint density of the vectors $\mathbf{y}_1$ and $\mathbf{y}_2$, conditional on τ, is the following $n_1 + n_2$ bivariate normal distribution

$$\mathcal{N}_{n_1+n_2}\left(\begin{pmatrix} \mathbf{y}_1 \\ \mathbf{y}_2 \end{pmatrix} \middle| \begin{pmatrix} \mathbf{A}\mathbf{m} + \mathbf{b} \\ \mathbf{m} \end{pmatrix}, \tau \begin{pmatrix} \mathbf{S} + \mathbf{A}\mathbf{V}\mathbf{A}^\top & \mathbf{A}\mathbf{V} \\ \mathbf{A}\mathbf{A}^\top & \mathbf{V} \end{pmatrix} \right).$$

Then, the marginal distribution of $\mathbf{y}_1$ conditional on τ, is the n_1 variate normal distribution

$$\mathcal{N}_{n_1}(\mathbf{y}_1 | \mathbf{A}\mathbf{m} + \mathbf{b}, \tau(\mathbf{S} + \mathbf{A}\mathbf{V}\mathbf{A}^\top)).$$

Further, since the marginal distribution of τ is an $\mathrm{InvGa}(\tau | \xi, \upsilon)$, Lemma 6.1 follows from the definition of the normal–inverted–gamma distribution. This proves the assertion. $\square$

Lemma 6.2. *If* $(\mathbf{y}, \tau)$ *follows the normal–inverted–gamma distribution* $\mathcal{N}_n InvGa(\mathbf{y}, \tau | \mathbf{m}, \mathbf{V}; \alpha, \beta)$, *then the marginal of the* n *dimensional vector* $\mathbf{y}$ *is the following multivariate Student* t *distribution*

$$\mathcal{T}_n\left(\mathbf{y}|\mathbf{m}, \frac{\beta}{\alpha}\mathbf{V}; 2\alpha\right).$$

Proof. The joint density of $(\mathbf{y}, \tau)$ is

$$\mathcal{N}_n InvGa(\mathbf{y}, \tau | \mathbf{m}, \mathbf{V}; \alpha, \beta) = \frac{|\mathbf{V}|^{-1/2}}{(2\pi)^{-n/2}} \frac{\beta}{\Gamma(\alpha)} \tau^{-(\alpha+n/2+1)}$$

$$\times \exp\left\{-\frac{1}{\tau}\left[\beta + (\mathbf{y}-\mathbf{m})^\top \mathbf{V}^{-1}(\mathbf{y}-\mathbf{m})\right]\right\}.$$

If we integrate out the variable τ in the joint density, we can then recognize the density function of an n variate Student t distribution with location parameter $\mathbf{m}$, scale matrix $(\beta/\alpha)\mathbf{V}$, and degree of freedom 2α. This proves the assertion. $\square$

From the posterior distribution in (6.16), and Lemma 6.1 and 6.2 we can derive the Bayesian predictive distribution of the effectiveness e, given the data, as follows. Let $\mathbf{x_e}$ be a generic vector of regressors for the effectiveness e. Then, the distribution of e is

$$f(e|\boldsymbol{\beta}, \tau^2, \mathbf{x_e}) = \mathcal{N}(e|\mathbf{x_e^\top}\boldsymbol{\beta}, \tau^2)$$

and the predictive distribution of e and τ is the distribution

$$f(e, \tau|\mathbf{x_e}) = \mathcal{N}InvGa(e, \tau|\widehat{\boldsymbol{\beta}}^\top \mathbf{x_e}, (1 + \mathbf{x_e}(\mathbf{X_e^\top X_e})^{-1}\mathbf{x_e^\top}); \frac{\nu}{2}, \frac{\nu}{2}s^2).$$

Now, from the properties of the normal–inverted–gamma distribution, the Bayesian predictive marginal distribution of e is the following Student t distribution

$$f(e|\mathbf{x_e}) = \text{Student}(e|\mathbf{x_e}\widehat{\boldsymbol{\beta}}, (1 + \mathbf{x_e}(\mathbf{X_e^\top X_e})^{-1}\mathbf{x_e^\top})s_e^2; \nu). \qquad (6.17)$$

In a similar way, we can derive the Bayesian predictive distribution of c. In order to stress the possible dependency of this conditional predictive distribution on the effectiveness e when the effectiveness is an influential regressor, we now write $\mathbf{x_c}(\mathbf{e})$ instead of $\mathbf{x_c}$ to make clear the possible dependence on e. The conditional distribution of c given $\mathbf{e}$ and a generic regressor $\mathbf{x_c}(\mathbf{e})$ is

$$f(c|\mathbf{e}, \boldsymbol{\alpha}, \sigma^2, \mathbf{x_c}(\mathbf{e})) = \mathcal{N}(c|\mathbf{x_c}(\mathbf{e})\boldsymbol{\alpha}, \sigma^2).$$

As before, from equation (6.15) and Lemma 6.1, the Bayesian predictive distribution of c and σ given e, $f(c, \sigma|\mathbf{x_c}(\mathbf{e}))$ is

$$\mathcal{N}\text{InvGa}(c, \sigma^2|\mathbf{x_c}(\mathbf{e})\widehat{\boldsymbol{\alpha}}, 1 + \mathbf{x_c}(\mathbf{e})(\mathbf{X_c^\top X_c})^{-1}\mathbf{x_c}(\mathbf{e})^\top; \frac{\nu'}{2}, \frac{\nu'}{2}s'^2).$$

Then, the Bayesian predictive of the cost c, conditional on e, the data, and the regressors $\mathbf{x}_c(\mathbf{e})$ is the following Student t distribution

$$f(c|e, \mathbf{x_c}(\mathbf{e})) = \text{Student}(c|\widehat{\boldsymbol{\alpha}}^\top \mathbf{x_c}(\mathbf{e}), (1 + \mathbf{x_c}(\mathbf{e})(\mathbf{X_c^\top X_c})^{-1}\mathbf{x_c}(\mathbf{e})^\top)s_c^2; \nu').$$
(6.18)

Thus, from equations (6.17) and (6.18) we have that the joint Bayesian predictive distribution of c and e, conditional on the data and the generic regressors $\mathbf{x}_e$ and $\mathbf{x}_c(\mathbf{e})$, $f(c, e|\mathbf{x_c}(\mathbf{e}), \mathbf{x_e})$ is

$$\text{Student}(c|\widehat{\boldsymbol{\alpha}}^\top \mathbf{x_c}(\mathbf{e}), (1 + \mathbf{x_c}(\mathbf{e})(\mathbf{X'_c X_c})^{-1}\mathbf{x_c}(\mathbf{e})^\top)s_c^2; \nu')$$
$$\times \text{Student}(e|\widehat{\boldsymbol{\beta}}^\top \mathbf{x_e}, (1 + \mathbf{x_e}(\mathbf{X_e^\top X_e})^{-1}\mathbf{x_e^\top})s_e^2; \nu). \quad (6.19)$$

Unfortunately, $f(z|R, \text{data}, \mathbf{X_c}, \mathbf{X_e}, \mathbf{x_c}, \mathbf{x_e})$, the Bayesian predictive distribution of the net benefit z, has no an explicit form. However, sampling from the predictive distribution of the net benefit z is straightforward: first, for each R and the covariate $\mathbf{x_e}$, we sample from $f(e|\mathbf{x_e})$ in equation (6.16), and then for the covariate $\mathbf{x_c}$ from $f(c|e, \mathbf{x_c})$ in equation (6.18). From these samples of e's and c's we immediately have a sample of the net benefit z for any value of R and the covariates $\mathbf{x_c}, \mathbf{x_e}$.

6.4.2 The case where c and e are independent

A particular interesting case is that where $\mathbf{x_c}(\mathbf{e})$ does not depend on e, so that c and e are independent and the predictive distribution of z is the convolution of two Student t distributions. That is, equation (6.18) simplifies to

$$f(c|\mathbf{e}, \mathbf{x_c}) = \text{Student}(c|\mathbf{x_c}\widehat{\boldsymbol{\alpha}}, (1 + \mathbf{x_c}(\mathbf{X_c^T X_c})^{-1}\mathbf{x_c^T})s_c^2; \nu') \qquad (6.20)$$

and, from Theorem 1 of Girón et al. (1999), it follows that the predictive distribution of z is the following Behrens–Fisher distribution

$$\begin{aligned} f(z|R, \mathbf{x_e}, \mathbf{x_c}) \quad \sim \quad & \text{BeF}(z|R\mathbf{x_e}\boldsymbol{\beta} - \mathbf{x_c}\boldsymbol{\alpha}, (1 + \mathbf{x_e}(\mathbf{X_e^T X_e})^{-1}\mathbf{x_e^T})s_e^2 R^2 \\ & + (1 + \mathbf{x_c}(\mathbf{X_c^T X_c})^{-1}\mathbf{x_c^T})s_c^2, \nu, \nu'; \phi) \qquad (6.21) \end{aligned}$$

where the angle $\phi \in [0, \pi/2]$ is such that

$$\tan^2(\phi) = \frac{(1 + \mathbf{x_e}(\mathbf{X_e^T X_e})^{-1}\mathbf{x_e^T})\, s_e^2}{(1 + \mathbf{x_c}(\mathbf{X_c^T X_c})^{-1}\mathbf{x_c^T})\, s_c^2} R^2.$$

Further, formulae (6.15) and (6.17) simplify due to the fact that the scale terms of the t densities can be simplified and we have that

$$1 + \mathbf{x_e}(\mathbf{X_e' X_e})^{-1}\mathbf{x_e^T} = 1 + h_{jj}^{x_e},$$

and

$$1 + \mathbf{x_c}(\mathbf{X_c^T X_c})^{-1}\mathbf{x_c^T} = 1 + h_{jj}^{x_c},$$

where $h_{jj}^{x_e}$ and $h_{jj}^{x_c}$ are the diagonal elements of the corresponding hat matrices $\mathbf{H}^{x_e} = \mathbf{X_e}(\mathbf{X_e^T X_e})^{-1}\mathbf{X_e^T}$ and $\mathbf{H}^{x_c} = \mathbf{X_c}(\mathbf{X_c^T X_c})^{-1}\mathbf{X_c^T}$, respectively.

6.4.3   The lognormal–normal case

The case where the cost follows a lognormal distribution and the effectiveness a normal distribution, is analogous to the preceding normal–normal case with the exception that now the costs are replaced by their logcosts in the computations. The predictive distribution of the effectiveness is exactly the same as that of formula (6.14), while the predictive of the costs, conditional on the effectiveness, the equivalent of formula (6.16), is now a logStudent t distribution. In this case the predictive distribution of the net benefit has no known form. However, simulation of the joint distribution of (c, e) is straightforward, and from this sample we can obtain samples of the posterior predictive distribution of the net benefit z for any value of the parameter R and the covariates.

An important difference with the normal–normal case is that in the lognormal–normal case the expectation of z does not exist because the logStudent t distribution has no mean. This means that we cannot use the $U_1(z|R)$ utility function. Fortunately, we can compute the optimal treatment for the $U_2(z|R)$ utility function.

6.5 Optimal treatments for subgroups

A way of presenting the optimal treatments, conditional on R and $\mathbf{x}$, is as follows. For simplicity in the presentation, we consider the case of two treatments T_1 and T_2 and denote z_1 as the net benefit of treatment T_1 and z_2 the net benefit of treatment T_2. For the utility function $U_1(z|R)$, and a fixed value of R, the set of covariate values, or equivalently the subgroups, for which treatment T_1 is optimal is given by

$$\mathfrak{C}_R^{U_1} = \Big\{ \mathbf{x} : \varphi(R, \mathbf{x}) \geq 0 \Big\},$$

where

$$\varphi(R, \mathbf{x}) = \mathbb{E}_{P_1}(z_1|R, \mathrm{data}_1, \mathbf{x}) - \mathbb{E}_{P_2}(z_2|R, \mathrm{data}_2, \mathbf{x}).$$

Likewise, assuming that z_i is continuous, $i = 1, 2$, when using the utility function $U_2(z_i|R)$ for fixed R, the set of covariate values, or equivalently the subgroup, for which T_1 is optimal is given by

$$\mathfrak{C}_R^{U_2} = \Big\{ \mathbf{x} : \psi(R, \mathbf{x}) \geq 1/2 \Big\},$$

where

$$\psi(R, \mathbf{x}) = \Pr(Z_1 \geq Z_2|R, \mathrm{data}_1, \mathrm{data}_2, \mathbf{x}).$$

The computational difficulty for characterizing $\mathfrak{C}_R^{U_1}$ and $\mathfrak{C}_R^{U_2}$ depends on the complexity of the functions $\varphi(R, \mathbf{x})$ and $\psi(R, \mathbf{x})$. When the components of $\mathbf{x}$ are discrete, the sets $\mathfrak{C}_R^{U_1}$ and $\mathfrak{C}_R^{U_2}$ are easily characterized by direct evaluation of the functions $\varphi(R, \mathbf{x})$ and $\psi(R, \mathbf{x})$.

6.6   Examples

We illustrate the theory developed in this chapter with two real data sets including a number of covariates. In the first example the cost and effectiveness are both assumed to be normal distributed, and in the second example the cost is lognormal and the effectiveness is normal. In Example 6.3 we compute and compare the subgroup optimal decisions under both the utility functions U_1 and U_2, conditional on R. In Example 6.4, the optimal decision under U_1 does not exist as the predictive distribution of the net benefit has very heavy tails, and thus we only compute the optimal decision for U_2, conditional on R.

To both sets of data we first apply the preceding Bayesian methodology to select the set of influential covariates for the effectiveness and cost of the treatments. This step is crucial for picking up the relevant subgroups, for which the optimal treatments are then found. We emphasize two important facts derived from the data: One is that the influential covariates for different treatments do not coincide, and the second is that the optimal treatment changes as the influential covariates change their value. The resulting optimal decisions are compared with those derived by assuming that there are no covariates involved in the problem.

Example 6.3. *The data in this example were obtained from a study to compare three methadone maintenance programs: high, medium and low intensity, for opioid–addicted patients (Puigdollers et al., 2003). A 12–month follow–up study of 586 patients beginning methadone treatment at five drug care centers in Barcelona was performed.*

We note that this example is an extension of Example 6.2. Although three methadone maintenance programs were compared in the original study, for simplicity of the illustration we only compare the medium–intensity program, to be denoted in the following as treatment T_1, and the high–intensity program, treatment T_2. To each program, 165 and 155 patients were randomly assigned, respectively. Table 6.4 shows the *mean and standard deviation for the effectiveness and cost of both treatments.*

TABLE 6.4

Summary of the effectiveness, costs and covariates in Example 6.3.

	T_1	T_2
Cost	645.27 (288.43)	725.70 (317.47)
Effectiveness	18.65 (22.25)	19.35 (23.09)
Sample size	$n_1 = 165$	$n_2 = 155$
Age	30.95 (6.12)	30.53 (6.43)
Years	10.53 (5.56)	10.17 (5.78)
HIV	0.24	0.30
Education	0.79	0.75
Poly–drug	0.28	0.52
NHP_0	41.28 (24.76)	42.47 (23.55)
Gender	0.21	0.28

After analyzing these data sets for normality, we found that a normal model might be appropriate for both the cost and the effectiveness of the two treatments. Thus, results in Section 6.4.1 will be used.

In addition, seven covariates of the patients were recorded in the study: age; years of illness; the existence of HIV; education, which takes a value 1 for patients who have at most elementary education, and 0 in all other cases; poly-drug use, which takes the value 1 for patients who use more than one psychoactive drug and 0 otherwise; the initial health status NHP_0; and the gender. Table 6.4 shows the mean and standard deviation of the observed covariates.

From the application of the objective Bayesian variable selection procedure to select the set of most influential covariates for the effectiveness and the cost of each treatment, we find that the cost does not depend on the effectiveness or on the rest of the covariates. Further, the variable selection procedure demonstrates that only three covariates are worthwhile taking into account for the effectiveness of treatment T_1 and *only one for treatment T_2. That is, the effectiveness of T_1 only depends on the covariates NHP_0, education, and gender. The Bayesian estimation of the regressors' coefficients in expressions (6.3) and (6.4), the posterior mean using a non–informative prior, are 0.55 (0.05 for the standard deviation), $-7.11(3.31)$, and $-12.39(3.29)$, respectively. The effectiveness of T_2 only depends on NHP_0 and the coefficient is estimated as 0.59 with a standard deviation of 0.06. In both cases, the set of most influential covariates has been selected with a posterior probability larger than 0.8.*

Computing optimal treatments

For comparison purposes we first compute the optimal treatments when no covariates are considered. The cut–off point in R of the equation $\varphi(R) = \mathbb{E}(z_1 - z_2 \geq 0 | R, data) = 0$ is $R = 90$, so that under U_1, treatment T_1 is the optimal one for values $R \leq 90$, and treatment T_2 for $R \geq 90$. Likewise, the solution to the equation $\psi(R) = \Pr(z_1 - z_2 \geq 0 | R, data) = 1/2$ is $R = 90$, so the optimal decisions under U_1 and U_2 coincide. The plots of the straight line $\varphi(R)$ and the curve $\psi(R)$ are shown in Figure 6.1.

From the curve $\psi(R)$ we learn that for any $R \geq 90$ the probability that the net benefit of T_2 is larger than that of T_1 by only 0.51, a weak statement on the strength of the preference of T_2 over T_1.

We now compute the optimal treatments in the presence of covariates. As we have seen, the variable selection procedure concludes that there are only three influential covariates. When using the utility function U_1 we plot in the plane (NHP_0, R) the region where T_i is optimal, $i = 1, 2$, which follows from the values of the function for each $a, b \in \{0, 1\}$,

$$\varphi(R, data, NHP_0, a, b)$$
$$= \mathbb{E}(z_1 - z_2 \geq 0 | R, data, NHP_0, education = a, gender = b),$$

and similarly when using the utility function U_2, which follows from the values of the function

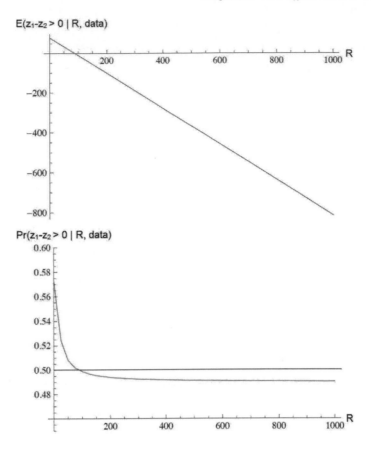

FIGURE 6.1

Graphics of $\varphi(R)$ (upper panel) and $\psi(R)$ (lower panel) in Example 6.3.

$\psi(R, data, NHP_0, a, b)$

$\qquad = \Pr(z_1 - z_2 \geq 0 | R, data, NHP_0, education = a, gender = b).$

The resulting four graphics for $a, b \in \{0, 1\}$, are displayed in Figure 6.2.

The optimal regions from $\varphi(R, data, NHP_0, a, b) \geq 0$ and $\psi(R, data, NHP_0, a, b) \geq 1/2$, $a, b = 0, 1$, coincide, so the optimal treatments when using either U_1 or U_2 are the same.

We want to remark that although the optimal decisions produced by considering the utility functions U_1 and U_2 are the same, the three-dimensional plots of the surfaces $\varphi(R, data, NHP_0, a, b) \geq 0$ and

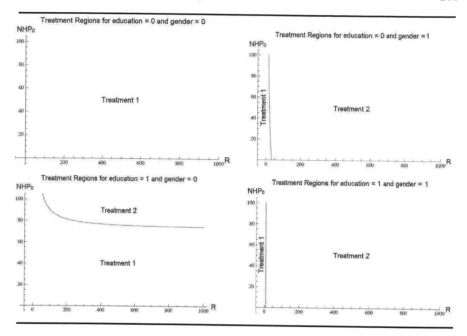

FIGURE 6.2

Regions for the allocation of the best treatment for the subgroups characterized by the three influential covariates. These regions are valid for the utility functions U_1 and U_2.

$\psi(R, data, NHP_0, a, b)$ for a, b fixed, are different, and they convey different information. In particular, the surface $\psi(R, data, NHP_0, a, b)$ for a, b fixed, provides more information on the strength of preference of a treatment over the other (measured by a probability), than the surface $\varphi(R, data, NHP_0, a, b)$ for a, b fixed, provides by the INB.

Figure 6.2 reveals the dramatic changes that the subgroup analysis produces on the allocation of any of the two treatments to the patients. The most striking conclusions we derive from Figure 6.2 are the following: For patients having the covariate education $= 0$, the gender is determinant for allocating the optimal treatment. For those having the covariate gender $= 0$, the optimal treatment is T_1 while for those having gender $= 1$ the optimal treatment is T_2, except for very small values of the parameter R; in fact, for $R < 25$.

The difference between the two graphics for the subgroups having gender = 1 is very small, and in this case the message is that treatment T_2 is the optimal one except for very small values of R.

For patients having education = 1 and gender = 0 the allocation of the best treatment depends heavily on the values of the covariate. For values of this covariate smaller than 74, the optimal treatment is T_1 while for larger values of this covariate the optimal treatment is T_2 except for small values of R, as indicated in the left-hand bottom panel.

We recall that the analysis, which does not take into account the co-variates, indicates that the optimal treatment is T_2, except for $R \leq 90$.

The conclusion from the cost–effectiveness analysis in this example is that the subgroup analysis adds information that might be crucial in the optimal allocation of a treatment.

Example 6.4. *This is an example based on real data from a randomized clinical trial (Hernández et al., 2003) that compares two alternative treatments for exacerbated chronic obstructive pulmonary disease (COPD) patients. It was postulated that home hospitalization, which is treatment T_2, of selected chronic obstructive pulmonary disease exacerbations admitted at the emergency room could facilitate a better outcome than conventional hospitalization, which is treatment T_1. For patients under treatment T_2, integrated care was delivered by a specialized respiratory nurse with the patient's free phone access to the nurse ensured for an 8–week follow–up period.*

We use information from 167 patients with COPD exacerbations over a 1–year period (1st November 1999 to 1st November 2000) among those admitted to the emergency department of two tertiary hospitals, Hospital Clínic and Hospital de Bellvitge of Barcelona, Spain. The two primary criteria for inclusion in the study were COPD exacerbation as a major cause of referral to the emergency room and absence of any criteria for imperative hospitalization as stated by the British Thoracic Society guidelines. The number of patients randomly allocated to treatment T_1 was 70 and 97 to T_2.

TABLE 6.5

Summary of the effectiveness, costs and covariates in Example 6.4.

	T_1	T_2
log(cost)	7.08 (1.11)	6.55 (0.98)
Effectiveness	−1.593 (20.15)	7.01 (14.07)
Sample size	$n_1 = 70$	$n_2 = 97$
Age	70.443 (9.22)	71.28 (9.90)
Gender	0.971	0.979
Smoker	0.171	0.268
FEV	0.398	0.434
HOSV	0.8715 (1.34)	0.567 (0.83)
$SGRQ_1$	50.121 (28.01)	54.986 (20.2)

The effectiveness of a treatment was the quality of life of the pa-
tients under the treatment measured by the difference between the score
at the beginning and at the end of the study on the St. George's Res-
piratory Questionnaire (SGRQ). SGRQ is one of the most widely used
instruments for assessing health–related quality of life in respiratory
patients (Ferrer et al., 2002). This variable SGRQ ranges from 0 to
100; a zero score indicates no impairment of overall health. The total
direct cost per patient, expressed in euros, includes hospital cost, am-
bulatory cost, pharmaceutical, and health–care costs. It can be checked
that the normal distribution fits the effectiveness dataset, and the log-
normal distribution fits the cost dataset.

The potential set of covariates considered includes age, gender,
smoking habit, forced expiratory volume in one second (FEV), exac-
erbations requiring in–hospital admission (HOSV), and the score at
the beginning of the study ($SGRQ_1$). Table 6.5 summarizes the dataset
(mean and standard deviation) of each treatment.

The results of applying the variable selection procedure to the ef-
fectiveness conclude that the influential covariates for the effectiveness
of treatment T_1 are age and $SGRQ_1$, and for the effectiveness of treat-
ment T_2 are FEV and $SGRQ_1$. The cost of treatment T_1 only has one

influential covariate, the FEV, while the influential covariates for the cost of treatment T_2 are age, $SGRQ_1$ and effectiveness (e_2). In passing, we note that the influential covariates change when the treatment changes.

Computing optimal treatments

Note that, in contrast with the cost–effectiveness analysis of the data in the preceding Example 6.3, we cannot establish comparisons between the analysis using the U_1 and U_2 utility functions as the expectations of U_1 do not exist for the Bayesian predictive for the lognormal–normal model.

For comparison purposes we first carry out a cost–effectiveness analysis of the treatments T_1 and T_2 for the utility function U_2 without the consideration of covariates. Figure 6.3 illustrates this cost–effectiveness analysis by showing the curve $\psi(R) = \Pr(z_1 - z_2 \geq 0 | R, data)$.

In Figure 6.3 we observe that the posterior probability $\psi(R)$ is smaller than 0.5 for any value of R, and hence the conclusion we get

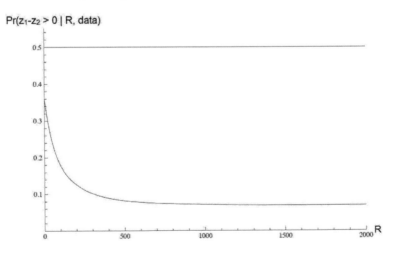

FIGURE 6.3

Graphic of the posterior probability $\Pr(z_1 - z_2 \geq 0 | R, \text{data})$, as a function of R.

*is that treatment T_2 is uniformly better than treatment T_1. This asser-
tion coincides with the conclusion reported in Hernández et al. (2003):
"The home hospitalization intervention generates better outcomes at
lower costs than conventional care."*

*However, if the cost–effectiveness analysis is carried out in the pres-
ence of the influential covariates, the above conclusion cannot be main-
tained, and now the optimality of treatment T_2 depends on the covariate values (or equivalently, depends on the type of patients). To illustrate this assertion, Figure 6.4 shows nine graphics in two dimensions
$(R, SGRQ_1)$ showing how the covariates FEV and age modify the treatment to be chosen. We recall that treatment T_1 is chosen when $\Pr(z_1 -
z_2 > 0 | R, data, SGRQ_1, FEV, age) \geq 0.5$, and treatment T_2 otherwise.*

*From Figure 6.4 it follows that the regions where T_1 is the optimal
treatment tends to be smaller as the age increases from 40 to 90, and
they are nearly empty regions for age ≥ 90. We note that this is true
for any value of FEV.*

*On the other hand, for young patients, age ≤ 40, the regions where
T_1 is optimal become smaller as FEV increases. The same happens
for intermediate ages. We also note that for older patients, T_2 is the
optimal treatment regardless the value of any other covariate.*

*The overall conclusion we can draw from this example is that the
information provided by the influential covariates can be crucial for
detecting the optimal treatment.*

6.7   Improving subgroup definition

The linear models utilized so far to model cost and effectiveness were
denoted as $P(c, e | \theta_i, \mathbf{x})$, where θ_i is an unknown multidimensional
parameter and $\mathbf{x}$ a covariate vector. This bidimensional model was
decomposed as $P(c, e | \theta_i, \mathbf{x}) = P(c | e, \theta_i, \mathbf{x}) P(e | \theta_i, \mathbf{x})$. This decomposi-
tion is certainly correct and it gives us a warning that an incoherency

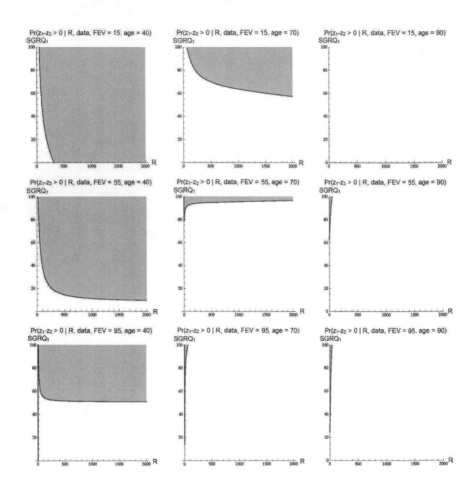

FIGURE 6.4

Regions indicating the optimal treatments in the plane (R, SGRQ_1) for some specific values of (FEV, age). Shaded areas represent the regions where the optimal treatment is T_1 and for the non–shaded regions the optimal treatment is T_2.

might appear when we separately apply the variable selection to the one-dimensional models for c and e.

Indeed, let us select covariates in the bidimensional model $P(c, e|\theta_i, \mathbf{x})$, the univariate conditional model $P(c|e, \theta_i, \mathbf{x})$, and the univariate marginal model $P(e|\theta_i, \mathbf{x})$, and let $\mathbf{x}_{sb}, \mathbf{x}_{sc}, \mathbf{x}_{se}$ be the resulting subsets of selected covariates for each of the models. Then, the models for the selected regressors $P(c, e|\theta_i, \mathbf{x}_{sb})$, $P(c|e, \theta_i, \mathbf{x}_{sc})$, and $P(e|\theta_i, \mathbf{x}_{se})$ do not satisfy the equality

$$P(c, e|\theta_i, \mathbf{x}_{sb}) = P(c|e, \theta_i, \mathbf{x}_{sc})P(e|\theta_i, \mathbf{x}_{se})$$

unless the three subsets of covariates are either equal or x_{sb} is the union of x_{sc} and x_{se}.

Furthermore, even when the three subsets of covariates coincide it is not clear how the underlying uncertainty in choosing the covariates in the marginal density of the effectiveness is propagated to the conditional distribution of the cost, and hence it is not clear how to compute the total variable selection uncertainty. We note that the model space for the effectiveness, in which model selection is carried out, contains 2^p models while the model space of the cost, conditional on the effectiveness, contains 2^{p+1} models. Therefore, our uncertainties are set in different probability spaces making it difficult to evaluate the total uncertainty in the statistical variable selection procedure. This has undesirable implications. For instance, if more than one model has a high posterior probability, inferences based on model averaging cannot be considered when the covariates are selected from the univariate decomposition because the weights are not well defined.

We remark that the subsets of covariates $\mathbf{x}_{sb}, \mathbf{x}_{sc}$ and $\mathbf{x}_{se}$ do not necessarily coincide as the following example shows. This example is based on a real clinical trial carried out in Hospital Clinic and Hospital de Bellvitge of Barcelona, Spain.

Example 6.4 (continued). *The Bayesian variable selection procedure for the intrinsic priors was applied to the data for each treatment and the bivariate model for the cost and the effectiveness, the cost model*

conditional on the effectiveness, and the model for the effectiveness.
The variables selected are given in Table 6.6.

TABLE 6.6
Influential variables for models of treatments T_1 and T_2.

Model	Treatment T_1	Treatment T_2
Bivariate model	{SGRQ1, FEV, age}	{SGRQ1, FEV}
Model for $c|e$	{ FEV }	{SGRQ1, age}
Model for e	{ SGRQ1, age}	{SGRQ1, FEV}

The conclusion we draw for treatment T_1 is that both ways of se-
lecting variables convey the same message, and hence no incoherency
appears. However, for treatment T_2 the influential covariates for the
bivariate model do not coincide with the ones selected when using the
univariate models.

A way to avoid the difficulty observed in the above example is to
return to the original bivariate distribution $P(c, e | \theta_i, \mathbf{x})$ and select the
covariates directly from this model. In this setting we have only one
model space containing 2^p bivariate models and the above–mentioned
difficulty disappears. The price we pay for defining subgroups with
the variables selected for the bivariate model is the higher complexity
of the underlying sampling model that now becomes a 1×2 matrix–
variate normal (or lognormal) distribution. Given the complexity of
this bivariate formulation, a slightly different notation with respect the
above is now introduced. A detailed study of this problem is given in
Moreno et al. (2013b).

Let us consider a sample of joint cost and effectiveness $\{\mathbf{y}_j = (c_j, e_j)^\top, \ j = 1, \ldots, n\}$ from n patients receiving a generic treatment
T, which we write as the $n \times 2$ matrix

$$\mathbf{Y} = \begin{pmatrix} c_1 & e_1 \\ \vdots & \vdots \\ c_n & e_n \end{pmatrix}. \tag{6.22}$$

Let $\mathbf{X}_p$ be the $n \times p$ design matrix

$$\mathbf{X}_p = \begin{pmatrix} x_{11} & \cdots & x_{1p} \\ \vdots & \ddots & \vdots \\ x_{n1} & \cdots & x_{np} \end{pmatrix}, \tag{6.23}$$

and $\mathbf{B}_p$ the $p \times 2$ matrix of regression coefficients

$$\mathbf{B}_p = \begin{pmatrix} \beta_{11} & \beta_{12} \\ \vdots & \vdots \\ \beta_{p1} & \beta_{p2} \end{pmatrix}. \tag{6.24}$$

The likelihood of the parameters $(\mathbf{B}_p, \mathbf{\Sigma}_p)$ for the data $(\mathbf{Y}, \mathbf{X}_p)$ is assumed to be given by the sampling matrix normal density

$$\mathcal{N}_{n \times 2} \left(\mathbf{Y} \mid \mathbf{X}_p \mathbf{B}_p, \sigma_p^2 (\mathbf{I}_n \otimes \mathbf{V}) \right), \tag{6.25}$$

where σ_p is an unknown positive number and $\mathbf{V}$ a 2×2 specified symmetric matrix.

Therefore, the sampling distribution of a submodel M_j containing a subset of j regressors, is given by the matrix–normal density $\mathcal{N}_{n \times 2} \left(\mathbf{Y} \mid \mathbf{X}_j \mathbf{B}_j, \sigma_j^2 (\mathbf{I}_n \otimes \mathbf{V}) \right)$, where $\mathbf{X}_j$ is the $n \times j$ design matrix resulting from suppressing the $p - j$ columns in $\mathbf{X}_p$, and $\mathbf{B}_j$ an unknown parameter matrix of dimensions $j \times 2$, and thus, the sampling full model M_p containing all p covariates coincides with (6.25). The number of such submodels is 2^p. To ensure that the intercept–only model is contained in any model we assume that $x_{k1} = 1$ for $k = 1, ..., n$, so that the number of submodels is now 2^{p-1}.

From Section 6.3, assuming a prior $\pi_j(\mathbf{B}_j, \sigma_j)$ for the model parameters and the standard uniform objective prior for models $\pi(M_j)$, $j \geq 1$, the posterior probability of a generic model M_j is given by

$$\Pr(M_j \mid \mathbf{Y}, \mathbf{X}_j) = \frac{B_{10}(\mathbf{Y}, \mathbf{X}_j)}{\sum_{k=0}^{2^p} B_{k0}(\mathbf{Y}, \mathbf{X}_k)}, \tag{6.26}$$

where $B_{k0}(\mathbf{Y}, \mathbf{X}_k)$ is the Bayes factor to compare model M_k against the intercept–only model M_0, and it is written as

$$B_{k0}(\mathbf{Y}, \mathbf{X}_k) = \frac{\int \int \mathcal{N}_{n \times 2}\left(\mathbf{Y} | \mathbf{X}_k \mathbf{B}_k, \sigma_k^2\left(\mathbf{I}_n \otimes \mathbf{V}\right)\right) \pi_k(\mathbf{B}_k, \sigma_k)\, d\mathbf{B}_k\, d\sigma_k}{\int \int \mathcal{N}_{n \times 2}\left(\mathbf{Y} | \mathbf{X}_1 \mathbf{B}_1, \sigma_1^2\left(\mathbf{I}_n \otimes \mathbf{V}\right)\right) \pi_1(\mathbf{B}_1, \sigma_1)\, d\mathbf{B}_1\, d\sigma_1}.$$

(6.27)

When intrinsic priors are considered as prior π_k, the Bayes factor for intrinsic priors is given by (Torres–Ruiz et al., 2011)

$$B_{k0}(\mathbf{Y}, \mathbf{X}_k) = 2(k+1)^{k-1} \int_0^{\pi/2} \frac{\sin(\varphi)^{2(k-1)+1}(n + (k+1)\sin^2 \varphi)^{(n-k)}}{\cos(\varphi)^{-1}[(k+1)\sin^2 \varphi + n\mathcal{B}_{k0}]^{(n-1)}}\, d\varphi,$$

(6.28)

where $\mathcal{B}_{k1}$ is the statistic

$$\mathcal{B}_{k0} = \frac{\operatorname{trace}\left(\mathbf{H}_{\mathbf{X}_k} \mathbf{Y} \mathbf{V}^{-1} \mathbf{Y}^\top\right)}{\operatorname{trace}\left(\mathbf{H}_{\mathbf{X}_1} \mathbf{Y} \mathbf{V}^{-1} \mathbf{Y}^\top\right)},$$

and $\mathbf{H}_{\mathbf{X}} = \mathbf{I}_n - \mathbf{X}(\mathbf{X}^\top \mathbf{X})^{-1}\mathbf{X}^\top$.

The statistic $\mathcal{B}_{k0}$ contains all the sample information needed for comparing model M_k against M_0. We call attention to the simplicity of the Bayes factor in formula (6.28) that generalizes the Bayes factor arising in the unidimensional regression normal case (Moreno et al., 2003).

Following the methodology presented in this chapter, the next step consists of finding the posterior predictive distribution of the net benefit. Suppose that from the application of the variable selection procedure to the original regression model for the effectiveness and the cost, we end up with the matrix–variate regression model $\mathcal{N}_{n \times 2}\left(\mathbf{Y} | \mathbf{X}_s \mathbf{B}_s, \sigma^2(\mathbf{I}_n \otimes \mathbf{V})\right)$, where $\mathbf{Y}$ is an $n \times 2$ matrix as in (6.22), $\mathbf{X}_s$ is the design matrix of the selected regressors that has dimensions $n \times q$, where $q \leq p$, and $\mathbf{B}_s$ is the $q \times 2$ submatrix of $\mathbf{B}_p$ corresponding to the selected regressors. Using results in Moreno et al. (2012), the posterior predictive distribution of the pair $(c, e)^\top$ for a generic vector of influential covariates $\mathbf{x}$ turns out to be the following bivariate Student t distribution

$$(c, e)^\top | \mathbf{x}, \mathbf{X}_s, \mathbf{Y}, \mathbf{V} \sim \mathcal{T}_2((c, e)^\top | \widehat{\mathbf{B}}_s^\top \mathbf{x}, s^2(1 + \mathbf{x}^\top (\mathbf{X}_s^\top \mathbf{X}_s)^{-1}\mathbf{x})\mathbf{V}; \nu),$$

(6.29)

where $\widehat{\mathbf{B}}_s = (\mathbf{X}_s^\top \mathbf{X}_s)^{-1} \mathbf{X}_s^\top \mathbf{Y}$ is the least squares or maximum likelihood estimator of $\mathbf{B}_s$, $\nu = 2(n-p)$ is the number of degrees of freedom, $s^2 = \text{trace}(\mathbf{V}^{-1}\mathbf{R}^\top \mathbf{R})/\nu^2$ is the unbiased estimator of the variance σ^2, and $\mathbf{R} = \mathbf{Y} - \mathbf{X}_s \widehat{\mathbf{B}}_s$ is the matrix of residuals.

From the properties of the multivariate Student t distribution, the posterior predictive distribution of the net benefit $z = e\,R - c$ is the following univariate Student t distribution, i.e., $z|\mathbf{x}, \mathbf{X}_s, \mathbf{Y}, \mathbf{V}$ follows a

$$\text{Student}(z | (-1, R)\widehat{\mathbf{B}}_s^\top \mathbf{x}^\top, s^2(1 + \mathbf{x}^\top (\mathbf{X}_s^\top \mathbf{X}_s)^{-1} \mathbf{x})$$
$$\times ((-1, R)\mathbf{V}(-1, R)^\top); \nu). \tag{6.30}$$

This predictive distribution depends on the matrix $\mathbf{V}$ which has to be assessed or estimated from the data. One obvious way to estimate the covariance matrix of the error term $\mathbf{\Sigma} = \sigma^2 \mathbf{V}$ of a linear model is through the maximum likelihood estimator

$$\widehat{\mathbf{\Sigma}} = \frac{1}{n}\mathbf{R}^\top \mathbf{R},$$

or the unbiased estimator based on the maximum likelihood one

$$\widehat{\mathbf{\Sigma}} = \frac{1}{n-p}\mathbf{R}^\top \mathbf{R},$$

or another Bayesian estimator of $\mathbf{\Sigma}$ proportional to the matrix $\mathbf{R}^\top \mathbf{R}$.

Example 6.5. *This is a simulated example inspired by the case study developed in Section 4.7 in Chapter 4 where real data taken for four treatments from a clinical trial (Pinto et al., 2000) were analyzed. Data were collected for direct costs (pharmaceutical, medical visit, and diagnostic test costs), and the effectiveness was measured by quality–adjusted life years (QALYs). The QALYs were calculated as the area above/below the EuroQOL visual analogue scale (VAS) (Richardson and Manca, 2004). The VAS is a self–rating of health–related quality of life that simulates a thermometer, ranging from a minimum of 0 (the worst health state imaginable) to a maximum of 100 (the best one). All 477 patients were followed–up for 6 months.*

The potential set of covariates in the data are six: x_1 denotes age, x_2

sex, x_3 the existence of concomitant illnesses which takes a value 1 if a concomitant illness is observed and 0 in all other cases, x_4 takes value 1 if two or more concomitant illnesses exist and 0 in all other cases. The concomitant illnesses that were studied were hypertension, cardiovascular diseases, allergies, asthma, diabetes, gastrointestinal alterations, urinary troubles, prior renal disease, raised cholesterol and/or triglyceride levels, chronic skin problems, or depression/anxiety. Other potential covariates are x_5 denoting the time, in months, from the onset of illness to obtaining real clinical data, and x_6 represents the initial health status measured through the VAS at the beginning of the treatment (VAS).

TABLE 6.7

Summary of the simulated effectiveness, the cost (in euros) of four treatments (standard deviations in parenthesis), and the covariates in Example 6.5.

	Treatment			
	T_1	T_2	T_3	T_4
Effectiveness	0.429	0.433	0.421	0.393
	(0.11)	(0.115)	(0.112)	(0.141)
Cost	7083.09	7253.60	6268.45	5948.12
	(1662.09)	(1530.04)	(1719.61)	(1398.19)
Covariates				
x_1 (years)	35.26	33.95	34.70	33.64
	(7.36)	(6.77)	(8.68)	(9.73)
x_2 (female)	0.29	0.35	0.37	0.12
x_3	0.27	0.33	0.30	0.24
x_4	0.11	0.07	0.19	0.08
x_5	79.37	87.18	69.10	58.52
	(91.99)	(138.42)	(51.35)	(41.78)
x_6	76.89	78.07	74.70	74.20
	(16.60)	(16.84)	(16.46)	(17.60)
n	268	93	91	25

The file we have used for illustrating the subgroup analysis is based on the real design matrices whose summary is in Table 6.7, where we have simulated the cost and effectiveness from a bivariate normal linear model keeping the original design matrices. Thus, for each treatment $i = 1, 2$, the data $\mathbf{Y}_i$ come from the model

$$\mathcal{N}_{n \times 2}\left(\mathbf{Y}_i \middle| \mathbf{X}_i \mathbf{B}_i, \mathbf{I}_n \otimes \boldsymbol{\Sigma}_i\right),$$

where $\mathbf{B}_i$ was chosen as a 6×2 matrix whose coefficients were the same as the least squares estimation of the regression matrix with the original data, and the covariance matrix $\boldsymbol{\Sigma}_i$ was chosen equal to the estimated covariance matrix. In this way, we reproduce a set of cost–effectiveness data which have similar values to the original ones but have the property of being bivariate normal distributed.

After applying the variable selection procedure, the models having the highest posterior probabilities for each of the treatments are given in Table 6.8.

TABLE 6.8

Model with the highest posterior model probability for each treatment in Example 6.5.

Treatment	Top model	Post. Prob.
T_1	$\{x_4, x_6\}$	0.923
T_2	$\{x_4\}$	0.778
T_3	$\{x_4, x_5, x_6\}$	0.437
T_4	$\{x_4\}$	0.511

In Figure 6.5 we present, in the plane defined by (R, x_6), optimal treatments for subgroups using the utility function U_1. For the subgroup defined by $(x_4, x_5) = (0, 0)$ the optimal treatments in the plane (R, x_6) are presented in the left upper panel: in the medium grey region, T_2 is optimal, in the dark grey, T_3, and in the light grey, T_4. For this subgroup, treatment T_1 is not optimal in any point of the plane. The interpretation of any other panel is similar. It is interesting to note *that for the subgroup* $(x_4, x_5) = (1,1)$ *treatment* T_3 *is the optimal one in almost any point in the plane.*

   The optimal treatments are located in the plane (R, x_6) *for all possible values of the covariates* x_4 *and* x_5, *and they are displayed in the four graphics in Figure 6.5.*

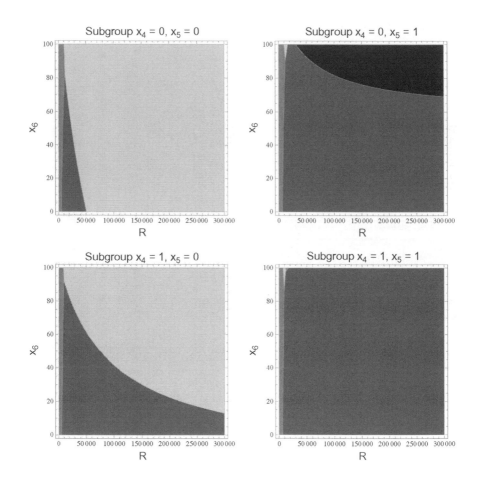

FIGURE 6.5

Regions indicating the optimal treatments in the plane (R, x_6) for the subgroup defined by (x_4, x_5) in Example 6.5. Colored areas represent the regions where the optimal treatments are T_1 (Black), T_2 (medium grey), T_3 (dark grey), and T_4 (light grey).

For comparative purposes, we present the optimal treatment as a function of R when no covariate is considered. We find that T_1 is never optimal, T_2 is optimal if $R \in (102382, \infty)$, T_3 is optimal if $R \in (11184, 102382)$, and T_4 is optimal if $R \in (0, 11184)$.

Example 6.4 (continued). *We saw in Table 6.6 that the influential covariates for the cost and effectiveness of treatment T_1 are $\{SGRQ1, FEV, age\}$ and for treatment T_2, $\{SGRQ1, FEV\}$.*

Figure 6.6 presents twelve plots illustrating the optimal treatment in the plane $(R, SGRQ_1)$ for subgroups defined by the covariates FEV and age and the utility function U_2. We recall that for this utility function, treatment T_1 is optimal when $\Pr(z_1 - z_2 \geq 0 | R, data, SGRQ_1, FEV, age) \geq 0.5$, and treatment T_2 otherwise.

The plots in Figure 6.6 essentially show that for patients with age below 50 years, the predominant treatment in the plane $(R, SGRQ_1)$ is T_1 and this situation reverses as the patients get older. For instance, for patients older than 80, the optimal treatment is always T_2, regardless of the covariate SGRQ1. For patients older than 80, these conclusions are close to the ones obtained in Example 6.4 in Section 6.6 where we used the decomposition of the bivariate lognormal–normal model into a product of two univariate models.

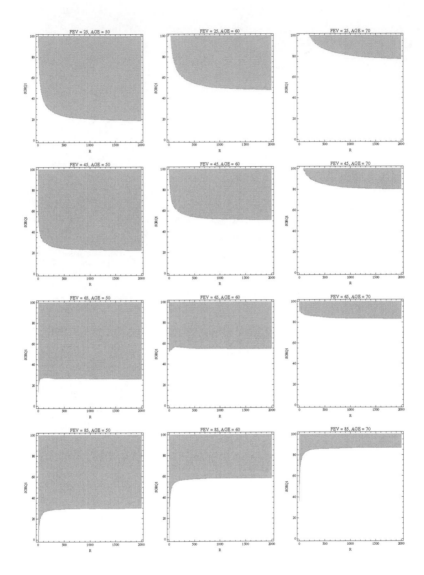

FIGURE 6.6

Regions indicating the optimal treatments in the plane $(R, \mathrm{SGRQ_1})$ for subgroups defined by the covariates (FEV, age) in Example 6.4 (continued). Shaded areas represent the regions where the optimal treatment is T_1 and for the non–shaded regions the optimal treatment is T_2.

Bibliography

M.J. Al and B.A. Van Hout. Bayesian approach to economic analysis of clinical trials: The case of stenting versus balloon angioplasty. *Health Economics*, 9(7):599–609, 2000.

M. Allais. Le comportement de l'homme rationnel devant le risque: Critique des postulats et axiomes de l'ecole americaine. *Econometrica*, 21(4):503–546, 1953.

J. Alonso, L. Prieto, and J.M. Anto. The Spanish version of the Nottingham Health Profile: A review of adaptation and instrument characteristics. *Quality of Life Research*, 3(6):385–393, 1994.

X. Badía, M. Roset, S. Montserrat, M. Herdman, and S. Segura. The Spanish version of EuroQol: A description and its applications. European Quality of Life scale. *Medicina Clínica*, 112:79–86, 1999.

G. Baio. *Bayesian Methods in Health Economics*. Chapman & Hall/CRC Biostatistics Series. Taylor & Francis, 2012.

G. Baio. Bayesian models for cost–effectiveness analysis in the presence of structural zero costs. *Statistics in Medicine*, 33(11):1900–1913, 2014.

G. Baio, A. Berardi, and A. Heath. *Bayesian Cost–Effectiveness Analysis with the R Package BCEA*. Cham: Springer, 2017.

D. Barry and J. A. Hartigan. Product partition models for change point problems. *Annals of Statistics*, 20(1):260–279, 1992.

M.S. Bartlett. A comment on D. V. Lindley's statistical paradox. *Biometrika*, 44(3/4):533–534, 1957.

T. Bayes. An essay towards solving a problem in the doctrine of chances. *Philosophical Transactions of the Royal Society of London*, 53:370–418, 1763.

I. Bebu, T. Mathew, and J.M. Lachin. Probabilistic measures of cost–effectiveness. *Statistics in Medicine*, 35(22):3976–3986, 2016.

J.O. Berger. *Statistical Decision Theory and Bayesian Analysis.* Springer, New York, 1985.

J.O. Berger. An overview of robust Bayesian analysis. *Test*, 3(1):5–124, 1994.

J.O. Berger and L.R. Pericchi. The intrinsic Bayes factor for model selection and prediction. *Journal of the American Statistical Association*, 91(433):109–122, 1996a.

J.O. Berger and L.R. Pericchi. The intrinsic Bayes factor for linear models (with discussion). In J.M. Bernardo, J.O. Berger, A.P. Dawid, and A.F.M. Smith, editors, *Bayesian Statistics 5*, pages 25–44. Oxford University Press, New York, 1996b.

J.O. Berger and L.R. Pericchi. Objective Bayesian methods for model selection: Introduction and comparison. In P. Lahiri, editor, *Model Selection*, volume 38 of *Lecture Notes–Monograph Series*, pages 135–207. Institute of Mathematical Statistics, Beachwood, OH, 2001.

J.O. Berger, B. Betró, E. Moreno, L.R. Pericchi, F. Ruggeri, G. Salinetti, and L. Wasserman, editors. *Bayesian Robustness*, volume 29 of *Lectures Notes–Monograph Series*. Institute of Mathematical Statistics, Hayward, CA, 1996.

J.O. Berger, J.M. Bernardo, and D. Sun. The formal definition of reference priors. *The Annals of Statistics*, 37(2):905–938, 2009.

J.O. Berger, M.J. Bayarri, and L.R. Pericchi. The effective sample size. *Econometric Reviews*, 33(1–4):197–217, 2014.

J.M. Bernardo. Reference posterior distributions for Bayesian inference. *Journal of the Royal Statistical Society. Series B (Methodological)*, 41(2):113–147, 1979.

J.M. Bernardo and A.F. M. Smith. *Bayesian Theory*. John Wiley & Sons, Inc., Chichester, 1994.

D.K. Bhaumik, A. Amatya, S.T. Normand, J. Greenhous, E. Kaizar, B. Neelon, and R.D. Gibbons. Meta–analysis of rare binary adverse event data. *Journal of the American Statistical Association*, 107 (498):555–567, 2012.

J. Bjøner and H. Keiding. Cost–effectiveness with multiple outcomes. *Health Economics*, 13(12):1181–1190, 2004.

A. Bobinac, J. van Exel, F.H. Rutten, and W.B.F. Brouwer. The value of a QALY: Individual willingness to pay for health gains under risk. *PharmacoEconomics*, 32(1):75–86, 2014.

A. Briggs, M. Sculpher, and K. Claxton. *Decision Modelling for Health Economic Evaluation*. Oxford University Press, 2006.

A.H. Briggs. Bayesian approach to stochastic cost–effectiveness analysis. *Health Economics*, 8(3):257–261, 1999.

A.H. Briggs. Handling uncertainty in cost–effectiveness models. *Pharmacoeconomics*, 17(5):479–500, 2000.

A.H. Briggs and P. Fenn. Confidence intervals or surfaces? Uncertainty on the cost–effectiveness plane. *Health Economics*, 7(8):723–740, 1998.

A.H. Briggs and A.M. Gray. Handling uncertainty when performing economic evaluation of health care interventions. *Health Technology Assessment*, 3(2):1–134, 1999.

A.H. Briggs, D.E. Wonderling, and C.Z. Mooney. Pulling cost–effectiveness analysis up by its bootstrap: A non–parametric approach to confidence interval estimation. *Health Economics*, 6(4): 327–340, 1997.

R. Brooks. EuroQol: The current state of play. *Health Policy*, 37(1): 53–72, 1996.

J.M. Brophy and L. Joseph. Placing trials in context using Bayesian analysis: GUSTO revisited by Reverend Bayes. *Journal of the American Medical Association*, 273(1):871–875, 1995.

T. Burns, F. Creed, T. Fahy, S. Thompson, P. Tyrer, and I. White. Intensive versus standard case management for severe psychotic illness: A randomised trial. UK 700 Group. *Lancet*, 353(9171):2185–2189, 1999.

P.L. Canner. An overview of six clinical trials of aspirin in coronary heart disease. *Statistics in Medicine*, 6(3):255–263, 1987.

G. Casella and E. Moreno. Objective Bayesian variable selection. *Journal of the American Statistical Association*, 101(473):157–167, 2006.

G. Casella and E. Moreno. Assessing robustness of intrinsic tests of independence in two–way contingency tables. *Journal of the American Statistical Association*, 104(487):1261–1271, 2009.

G. Casella, E. Moreno, and F. J. Girón. Cluster analysis, model selection, and prior distributions on models. *Bayesian Analysis*, 9(3): 613–658, 2014.

K. Chaloner. Elicitation of priors distributions. In D.A. Berry and D.K. Stangl, editors, *Bayesian Biostatistics*. Marcel Dekker, New York, 1996.

K. Chaloner and F.S. Rhame. Quantifying and documenting prior beliefs in clinical trials. *Statistics in Medicine*, 20(4):581–600, 2001.

M.A. Chaudhary and S.C. Stearns. Estimating confidence intervals for cost–effectiveness ratios: An example from a randomized trial. *Statistics in Medicine*, 15(13):1447–1458, 1996.

S. Chib and L. Jacobi. Modeling and calculating the effect of treatment at baseline from panel outcomes. *Journal of Econometrics*, 140(2): 781–801, 2007.

O. Ciani and C. Jommi. The role of health technology assessment bodies in shaping drug development. *Drug Design, Development and Therapy*, 8:2273–2281, 2014.

M. Clyde and E.I. George. Model uncertainty. *Statistical Science*, 19 (1):81–94, 2004.

M. Clyde, G. Parmigiani, and B. Vidakovic. Multiple shrinkage and subset selection in wavelets. *Biometrika*, 85(2):391–401, 1998.

M. Collins and N. Latimer. Nice's end of life decision making scheme: Impact on population health. *British Medical Journal*, 346:f1363, 2013.

G. Consonni and L. La Rocca. Tests based on intrinsic priors for the equality of two correlated proportions. *Journal of the American Statistical Association*, 103(483):1260–1269, 2008.

G. Consonni and P. Veronese. A Bayesian method for combining results from several binomial experiments. *Journal of the American Statistical Association*, 90(431):935–944, 1995.

G. Consonni, E. Moreno, and S. Venturini. Testing Hardy–Weinberg equilibrium: An objective Bayesian analysis. *Statistics in Medicine*, 30(1):62–74, 2011.

G. Consonni, J.J. Forster, and L. La Rocca. The Whetstone and the Alum Block: Balanced objective Bayesian comparison of nested models for discrete data. *Statistical Science*, 28(3):398–423, 08 2013.

G. Consonni, D. Fouskakis, B. Liseo, and I. Ntzoufras. Prior distributions for objective Bayesian analysis. *Bayesian Analysis*, 13(2): 627–679, 2018.

N.J. Cooper, D. Spiegelhalter, S. Bujkiewicz, P. Dequen, and A.J. Sutton. Use of implicit and explicit Bayesian methods in health technology assessment. *International Journal of Technology Assessment in Health Care*, 29(3):336–342, 2013.

J.M. Corcuera and F. Giummolè. A generalized Bayes rule for prediction. *Scandinavian Journal of Statistics*, 26(2):265–279, 1999.

J.E. Cornell, C.D. Mulrow, R. Localio, C.B. Stack, A.R. Meibohm, E. Guallar, and S.N. Goodman. Random–effects meta–analysis of inconsistent effects: A time for change. *Annals of Internal Medicine*, 160(4):267–270, 2014.

B. Cosmi, C. Legnani, M. Cini, E. Favaretto, and G. Palareti. D–dimer and factor VIII are independent risk factors for recurrence after anticoagulation withdrawal for a first idiopathic deep vein thrombosis. *Thrombosis Research*, 122(5):610–617, 2008.

E.M. Crowley. Product partition models for normal means. *Journal of the American Statistical Association*, 92(437):192–198, 1997.

C. Davies, A. Briggs, P. Lorgelly, G. Garellick, and H. Malchau. The "hazards" of extrapolating survival curves. *Medical Decision Making*, 33(3):369–380, 2013.

M.H. DeGroot. *Optimal Statistical Decisions*. McGraw–Hill, New York, 1970.

R. DerSimonian and N. Laird. Meta–analysis in clinical trials. *Controlled Clinical Trials*, 7(3):177–188, 1986.

S.A. Detsky and G.N. Naglie. A clinician's guide to cost–effectiveness analysis. *Annals of Internal Medicine*, 113(2):147–154, 1990.

P. Dolan. Thinking about it: Thoughts about health and valuing QALYs. *Health Economics*, 20(12):1407–1416, 2011.

C. Donalson, R. Baker, H. Mason, M. Jones–Lee, E. Lancsar, J. Wild-man, I. Bateman, G. Loomes, A. Robinson, R. Sugden, J.L. Pinto–Prades, M. Ryan, P. Shackley, and R. Smith. The social value of a QALY: Raising the bar or barring the raise? *BMC Health Services Research*, 11:8, 2011.

M.F. Drummond, G.L. Stoddart, and G.W. Torrance. *Methods for the Economic Evaluation of Health Care Programmes.* Oxford University Press, Oxford, 1987.

M.F. Drummond, H. Weatherly, and B. Ferguson. Economic evaluation of health interventions. *British Medical Journal*, 337:a1204, 2008.

M.F. Drummond, M.J. Sculpher, K. Claxton, G.L. Stoddart, and G.V. Torrance. *Methods for the Economic Evaluation of Health Care Programmes.* Oxford: Oxford University Press, 4th edition, 2015.

W. DuMouchel and C. Waternaux. Hierarchical models for combining information and for meta–analyses (with discussion). In J.M. Bernardo, J.O. Berger, A. Dawid, and A.F.M. Smith, editors, *Bayesian Statistics 4*, pages 338–341. Clarendon Press, Oxford, 1992.

S. Eckermann, A. Briggs, and A.R. Willan. Health technology assessment in the cost–disutility plane. *Medical Decision Making*, 28(2): 172–181, 2008.

D.M. Eddy, V. Hasselblad, and R.D. Schachter. *Meta–analysis by the Confidence Profile Method: The Statistical Synthesis of Evidence.* Academic Press, Boston, 1992.

B. Efron and R.J. Tibshirani. *An Introduction to the Bootstrap.* Chapman & Hall, New York, 1993.

J.M. Eisenberg. Clinical economics: A guide to economic analysis of clinical practice. *Journal of the American Medical Association*, 262 (20):2879–2886, 1989.

R. Ernst. Indirect costs and cost–effectiveness analysis. *Value in Health*, 9(4):253–261, 2006.

M.A. Espinoza, M. Andrea, K. Claxton, and M.J. Sculpher. The value of heterogeneity for cost–effectiveness subgroup analysis. *Medical Decision Making*, 34(8):951–964, 2014.

J. Evans, R. and J. Sedransk. Combining data from experiments that may be similar. *Biometrika*, 88(3):643–656, 2001.

E. Fenwick and S. Byford. A guide to cost–effectiveness acceptability curves. *British Journal of Psychiatry*, 187(2):106–108, 2005.

E. Fenwick, K. Claxton, and M. Sculpher. Representing uncertainty: The role of cost–effectiveness acceptability curves. *Health Economics*, 10(8):779–787, 2001.

E. Fenwick, B.J. O'Brien, and A. Briggs. Cost–effectiveness acceptability curves: Facts, fallacies and frequently asked questions. *Health Economics*, 13(5):405–415, 2004.

E. Fenwick, K. Claxton, and M. Sculpher. The value of implementation and the value of information: Combined and uneven development. *Medical Decision Making*, 28(1):21–32, 2008.

C. Fernández, E. Ley, and M. Steel. Benchmark priors for Bayesian model averaging. *Journal of Econometrics*, 100(2):381–427, 2001.

M. Ferrer, C. Villasante, J. Alonso, V. Sobradillo, R. Gabriel, G. Vilagut, J.F. Masa, J.L. Viejo, C.A. Jiménez-Ruiz, and M. Miravitlles. Interpretation of quality of life scores from the St George's Respiratory Questionnaire. *European Respiratory Journal*, 19(3):405–413, 2002.

E.C. Fieller. Some problems in interval estimation (with discussion). *Journal of the Royal Statistical Society, Series B*, 16(2):175–188, 1954.

R.A. Fisher. On the mathematical foundations of theoretical statistics. *Philosophical Transactions of the Royal Society of London, Series A*, 222:309–368, 1922.

L.S. Freedman and D.J. Spiegelhalter. The assessment of the subjective opinion and its use in relation to stopping rules for clinical trial. *Journal of the Royal Statistical Society. Series D (The Statistician)*, 33(1/2):153–160, 1983.

T. Friede, C. Röver, S. Wandel, and B. Neuenschwander. Meta–analysis of few small studies in orphan diseases. *Research Synthesis Methods*, 8(1):79–91, 2017.

D.G. Fryback, J.O. Chinnis, and J.W. Ulvila. Bayesian cost–effectiveness analysis. An example using the GUSTO trial. *International Journal of Technology Assessment in Health Care*, 17(1): 83–97, 2001.

A. Gelman, J. Carlin, H. Stern, and D. Rubin. *Bayesian Data Analysis (2nd edition)*. Chapman & Hall, London, 2004.

C. Genest and J. Zidek. Combining probability distributions: A critique and an annotated bibliography (with discussion). *Statistical Science*, 1(1):114–148, 1986.

E.I. George and D.P. Foster. Calibration and empirical Bayes variable selection. *Biometrika*, 87(4):731–747, 2000.

E.I. George and R.E. McCulloch. Variable selection via Gibbs sampling. *Journal of the American Statistical Association*, 88(423):881–889, 1993.

E.I. George and R.E. McCulloch. Approaches for Bayesian variable selection. *Statistica Sinica*, 7(2):339–373, 1997.

F.J. Girón, M.L. Martínez, and L. Imlahi. A characterization of the Behrens–Fisher distribution with applications to Bayesian inference.

Comptes Rendus de l'Académie des Sciences - Series I - Mathematics, 328(8):701–706, 1999.

F.J. Girón, E. Moreno, and M.L. Martínez. An objective Bayesian procedure for variable selection in regression. In N. Balakrishnan, J. M. Sarabia, and E. Castillo, editors, *Advances in Distribution Theory, Order Statistics, and Inference*, pages 389–404. Birkhäuser, Boston, MA, 2006.

F.J. Girón, E. Moreno, and G Casella. A Bayesian analysis of multiple change points in linear models (with discussion). In J.M. Bernardo, M.J. Bayarri, J.O. Berger, A.P. Dawid, D. Heckerman, A.F.M. Smith, and M West, editors, *Bayesian Statistics 8*, pages 227–252. Oxford University Press, New York, 2007.

M.R. Gold, J.E. Siegel, L.B. Russell, and M.C. Weinstein. *Cost–Effectiveness in Health and Medicine*. Oxford University Press, New York, 1996.

N. Gomes, E.S.W. Ng, R. Grieve, R. Nixon, J. Carpenter, and S.G. Thompson. Developing appropriate methods for cost–effectiveness analysis of cluster randomized trials. *Medical Decision Making*, 32 (2):350–361, 2012.

J.S. Goodwin, W.C. Hunt, C.G. Humble, C.R. Key, and J.M. Samet. Cancer treatment protocols. Who gets chosen? *Archives of Internal Medicine*, 148(1):2258–2260, 1988.

A. Gray, M. Raikou, A. McGuire, P. Fenn, R. Stevens, C. Cull, I. Stratton, A. Adler, R. Holman, and R. Turner. Cost effectiveness of an intensive blood glucose control policy in patients with type 2 diabetes: Economic analysis alongside randomised controlled trial (UKPDS 41). *British Medical Journal*, 320(7246):1373–1378, 2000.

R. Grieve, R. Nixon, S.G. Thompson, and C. Normand. Using multilevel models for assessing the variability of multinational resource use and cost data. *Health Economics*, 14(2):185–196, 2005.

M.H. Hansen and B. Yu. Model selection and the principle of minimum description length. *Journal of the American Statistical Association*, 96(454):746–774, 2001.

J.A. Hartigan. Partition models. *Communications in Statistics –Theory and Methods*, 19(8):2745–2756, 1990.

J. Hartung and G. Knapp. On tests of the overall treatment effect in meta–analysis with normally distributed responses. *Statistics in Medicine*, 20(12):1771–1782, 2001.

D. Hayley. The history of health technology assessment in Australia. *International Journal of Technology Assessment in Health Care*, 25 (Suppl 1):61–67, 2009.

D.F. Heitjan. Bayesian interim analysis of Phase II cancer clinical trials. *Statistics in Medicine*, 16(16):1791–1802, 1997.

D.F. Heitjan. Fieller's method and net health benefits. *Health Economics*, 9(4):327–335, 2000.

D.F. Heitjan, A.J. Moskowitz, and W. Whang. Bayesian estimation of cost–effectiveness ratios from clinical trials. *Health Economics*, 8(3): 191–201, 1999.

D. Henry. Economics analysis as an aid to subsidisation decisions: The development of Australia's guidelines for pharmaceuticals. *Pharmacoeconomics*, 1(1):54–67, 1992.

C. Hernández, A. Casas, J. Escarrabill, J. Alonso, J. Puig–Junoy, E. Farrero, G. Vilagut, B. Collvinent, R. Rodríguez–Rosin, and J. Roca. Home hospitalisation of exacerbated chronic obstructive pulmonary disease patients. *European Respiratory Journal*, 21(1): 58–67, 2003.

P. Hirskyj. QALY: An ethical issue that dare not speak its name. *Nursing Ethics*, 14(1):72–82, 2007.

R.A. Hirth, M.E. Chernew, E. Miller, M. Fendrick, and W.G. Weissert. Willingness to pay for a quality–adjusted life year. *Medical Decision Making*, 20(3):332–342, 2000.

J.S. Hoch, M.A. Rockx, and A.D. Krahn. Using the net benefit regression framework to construct cost–effectiveness acceptability curves: An example using data from a trial of external loop recorders versus Holter monitoring for ambulatory monitoring of "community acquired" syncope. *BMC Health Services Research*, 6(1):68, Jun 2006.

R.I. Horwitz, B.H. Singer, R.W. Makuch, and C.M. Viscoli. Can treatment that is helpful on average be harmful to some patients? A study of the conflicting information needs of clinical inquiry and drug regulation. *Journal of Clinical Epidemiology*, 49(4):395–400, 1996.

J. Hughes. Palliative care and the QALY problem. *Health Care Analysis*, 13(4):289–301, 2005.

C.P. Hunter, R.W. Frelick, A.R. Feldman, A.R. Bavier, W.H. Dunlap, L. Ford, D. Henson, D. MacFarlane, C.R. Smart, R. Yancik, and J.W. Yates. Selection factors in clinical trials: Results from the Community Clinical Oncology Program Physician's Patient Log. *Cancer Treatment Reports*, 71(6):559–565, 1987.

J. Hutton. Cost–benefit analysis in health care expenditure decision–making. *Health Economics*, 1(4):213–216, 1992.

M. Jakubczyk and B. Kaminski. Cost–effectiveness acceptability curves: Caveats quantified. *Health Economics*, 19(8):955–963, 2010.

H. Jeffreys. *The Theory of Probability*. Oxford Univ. Press, Oxford, 1961.

N.L. Johnson, Ad.W. Kemp, and S. Kotz. *Univariate Discrete Distributions*. John Wiley & Sons, Inc., New York, 2005.

D.A. Jones. Bayesian approach to the economic evaluation of health care technologies. In B. Spiker, editor, *Quality of Life and Phama-coeconomics in Clinical Trials*, pages 1189–1196. Lippincott–Raven, Philadelphia, 1996.

J.B. Kadane and L.J. Wolfson. Priors for the design and analysis of clinical trials. In D. Berry and D. Stangl, editors, *Bayesian Biostatistics*, pages 157–186. Marcel Dekker, New York, 1995.

R.M. Kaplan and J.W. Bush. Health–related quality of life measurement for evaluation research and policy analysis. *Health Psychology*, 1(1):61–80, 1982.

R.E. Kass and L. Wasserman. The selection of prior distributions by formal rules. *Journal of the American Statistical Association*, 91 (435):1343–1370, 1996.

M. Knapp and R. Mangalore. The trouble with QALYs. *Epidemiologia e Psichiatria Sociale*, 16(4):289–293, 2007.

B.G. Koerkamp, M.G.M. Hunink, T. Stijnen, J.K. Hammitt, K.M. Kuntz, and M.C. Weinstein. Limitations of acceptability curves for presenting uncertainty in cost–effectiveness analysis. *Medical Decision Making*, 27(2):101–111, 2007.

A.N. Kolmogorov. Determination of dispersion center and of accuracy measure from a finite number of observations. *Izvestija Akademii Nauk SSSR, Ser. Mat.*, 6:3–282, 1942.

F. Komaki. On asymptotic properties of predictive distributions. *Biometrika*, 83(2):299–313, 1996.

I.B. Korthals–de Bos, N. Smidt, M.W. van Tulder, M.P. Rutten–van Molken, H.J. Ader, D.A. van der Windt, W.J. Assendelft, and L.M. Bouter. Cost effectiveness of interventions for lateral epicondylitis: Results from a randomised controlled trial in primary care. *Pharmacoeconomics*, 22(3):183–195, 2004.

S. Kotz and S. Nadarajah. *Multivariate t–Distributions and Their Applications*. Cambridge University Press, 2004.

R.L. Kravitz, N. Duan, and J.L Braslow. Evidence–based medicine, heterogeneity of treatment effects, and the trouble with averages. *The Milbank Quarterly*, 82(4):661–687, 2004.

P. Lahiri. *Model Selection*. Institute of Mathematical Statistics, Beacwood, Ohio, 2001.

E.M. Laska, M. Meisner, and C. Siegel. Statistical inference for cost–effectiveness ratios. *Health Economics*, 6(3):229–242, 1997.

K.M. Lee and C.E. McCarron. *Guidelines for Economic Evaluation of Health Technologies: Canada*. Canadian Agency for Drugs and Technologies in Health (CADTH), Otawa, 2006.

E.L. Lehmann and G. Casella. *Theory of Point Estimation (Springer Texts in Statistics)*. Springer, 2nd edition, 1998.

L. León–Novelo, E. Moreno, and G. Casella. Objective Bayes model selection in probit models. *Statistics in Medicine*, 31(4):353–365, 2012.

F. Liang, R. Paulo, G. Molina, M.A. Clyde, and J.O. Berger. Mixtures of g–priors for Bayesian variable selection. *Journal of the American Statistical Association*, 103(481):410–423, 2008.

B. Liljas. How to calculate indirect costs in economic evaluations. *Pharmacoeconomics*, 13(1 Pt 1):1–7, 1998.

M. Loève. *Probability theory*. Van Nostrand, 1963.

M. Löthgren and N. Zethraeus. Definition, interpretation and calculation of cost–effectiveness acceptability curves. *Health Economics*, 9 (7):623–630, 2000.

B.R. Luce and R.E. Brown. The use of technology assessment by hospitals, health maintenance organizations, and third party payers in the United States. *International Journal of Technology Assessment in Health Care*, 11(1):79–92, 1995.

D. Malec and J. Sendrask. Bayesian methodology for combining the results from different experiments when the specifications for pooling are uncertain. *Biometrika*, 79(3):593–601, 1992.

A. Manca, N. Rice, M.J. Sculpher, and A.H. Briggs. Assessing generalisability by location in trial–based cost–effectiveness analysis: The use of multilevel models. *Health Economics*, 14(5):471–485, 2005.

A. Manca, P.C. Lambert, M. Sculpher, and N. Rice. Cost–effectiveness analysis using data from multinational trials: The use of bivariate hierarchical modeling. *Medical Decision Making*, 27(4):471–490, 2007.

J.S. Mandel, J.H. Bond, T.R. Church, D.C. Snover, G.M. Bradely, L.M. Schuman, and F. Ederer. Reducing mortality from colorectal cancer by screening for fecal occult blood. Minnesota Colon Cancer Control Study. *New England Journal of Medicine*, 328(19):1365–1371, 1993.

W.G. Manning, D.G. Fryback, and M.C. Weinstein. Reflecting uncertainty in cost–effectiveness analysis. In M.R. Gold, M.R. Siegel, L.B. Rusell, and M.C. Westein, editors, *Cost–Effectiveness in Health and Medicine*, pages 247–275. Oxford University Press, New York, 1996.

H. Mason, M. Jones–Lee, and C. Donaldson. Modelling the monetary value of a QALY: A new approach based on UK data. *Health Economics*, 18(3):933–950, 2009.

N. McCaffrey and S. Eckermann. Multiple effects cost–effectiveness analysis in cost–disutility space. In S. Eckermann, editor, *Health Economics from Theory to Practice*, pages 229–251. Adis International Lmited, New Zealand, 2016.

N. McCaffrey, M. Agar, J. Harlum, J. Karnon, D. Currow, and S. Eckermann. Better informing decision making with multiple outcomes cost–effectiveness analysis under uncertainty in cost–disutility space. *PLOS One*, 10(3):e0115544, 2015.

A. Mehrez and A. Gafni. Quality–adjusted life years, utility theory, and healthy–years equivalents. *Medical Decision Making*, 9(2):142–149, 1989.

E.J. Mishan. *Cost–Benefit Analysis*. Unwin Hyman, London, 1988.

E. Moreno. Bayes factors for intrinsic and fractional priors in nested models. Bayesian robustness. In Y. Dodge, editor, L_1–*Statistical Procedures and Related Topics*, volume 31 of *Lecture Notes–Monograph Series*, pages 257–270. Institute of Mathematical Statistics, Hayward, CA, 1997.

E. Moreno and F.J. Girón. Consistency of Bayes factors for intrinsic priors in normal linear models. *Comptes Rendus Mathématique*, 340 (12):911–914, 2005.

E. Moreno, F. Bertolino, and W. Racugno. An intrinsic limiting procedure for model selection and hypotheses testing. *Journal of the American Statistical Association*, 93(444):1451–1460, 1998.

E. Moreno, F.J. Girón, and F. Torres–Ruiz. Intrinsic priors for hypothesis testing in normal regression. *Revista de la Real Academia de Ciencias Exactas, Físicas y Naturales. Serie A. Matemáticas*, 97 (1):53–61, 2003.

E. Moreno, F.J. Girón, F.J. Vázquez–Polo, and M.A. Negrín. Optimal healthcare decisions: Comparing medical treatments on a cost–effectiveness basis. *European Journal of Operational Research*, 204 (1):180–187, 2010.

E. Moreno, F.J. Girón, F.J. Vázquez–Polo, and M.A. Negrín. Optimal healthcare decisions: The importance of the covariates in cost–effectiveness analysis. *European Journal of Operational Research*, 218(2):512–522, 2012.

E. Moreno, F.J. Girón, and A. García–Ferrer. A consistent on–line Bayesian procedure for detecting change points. *Environmetrics*, 24 (5):342–356, 2013a.

E. Moreno, F.J. Girón, M.L. Martínez, F.J. Vázquez–Polo, and M.A. Negrín. Optimal treatments in cost–effectiveness analysis in the presence of covariates: Improving patient subgroup definition. *European Journal of Operational Research*, 226(1):173–182, 2013b.

E. Moreno, F.J. Vázquez–Polo, and M.A. Negrín. Objective Bayesian meta–analysis for sparse discrete data. *Statistics in Medicine*, 33(21): 3676–3692, 2014.

E. Moreno, F.J. Girón, and G. Casella. Posterior model consistency in variable selection as the model dimension grows. *Statistical Science*, 30(2):228–241, 2015.

E. Moreno, F.J. Girón, and F.J. Vázquez–Polo. Cost effectiveness analysis for heterogeneous samples. *European Journal of Operational Research*, 254(1):127–137, 2016.

A. Murphy and R. Winkler. Reliability of subjective probability forecasts of precipitation and temperature. *Journal of the Royal Statistical Society. Series C (Applied Statistics)*, 26(1):41–47, 1977.

M.A. Negrín and F.J. Vázquez–Polo. Bayesian cost–effectiveness analysis with two measures of effectiveness: The cost–effectiveness acceptability plane. *Health Economics*, 15(4):363–372, 2006.

M.A. Negrín and F.J. Vázquez–Polo. Incorporating model uncertainty in cost–effectiveness analysis: A Bayesian model averaging approach. *Journal of Health Economics*, 27(5):1250 – 1259, 2008.

P. J. Neumann. Costing and perspective in published cost–effectiveness analysis. *Medical Care*, 47(7 Suppl 1):S28–S32, 2009.

NICE. Judging whether public health interventions offer value for money. Technical report, National Institute for Health and Care Excellence, 2013.

R.M. Nixon and S.G. Thompson. Methods for incorporating covariate adjustment, subgroup analysis and between–centre differences into cost–effectiveness evaluations. *Health Economics*, 14(12):1217–1229, 2005.

B.J. O'Brien, M.F. Drummond, R.J. Labelle, and A. Willan. In search of power and significance: Issues in the design and analysis of stochastic cost effectiveness studies in health care. *Medical Care*, 32(2): 150–163, 1994.

B.J. O'Brien, K. Gertsen, A.R. Willan, and A. Faulkner. Is there a kink in consumers' threshold value for cost–effectiveness in health care? *Health Economics*, 11(2):175–180, 2002.

A. O'Hagan and B.R. Luce. *A Primer on Bayesian Statistics in Health Economics and Outcomes Research*. Bayesian Initiative in Health Economics & Outcomes Research, Centre of Bayesian Statistics in Health Economics, Sheffield, 2003.

A. O'Hagan and J.W. Stevens. Bayesian assessment of sample size for clinical trials of cost–effectiveness. *Medical Decision Making*, 21(3): 219–230, 2001.

A. O'Hagan and J.W. Stevens. Bayesian methods for design and analysis of cost–effectiveness trials in the evaluation of health care technologies. *Statistical Methods in Medical Research*, 11(6):469–490, 2002.

A. O'Hagan and J.W. Stevens. Assessing and comparing costs: How robust are the bootstrap and methods based on asymptotic normality? *Health Economics*, 12(1):33–49, 2003.

A. O'Hagan, J.W. Stevens, and J. Montmartin. Inference for the cost–effectiveness acceptability curve and cost–effectiveness ratio. *Pharmacoeconomics*, 17(4):339–349, 2000.

A. O'Hagan, J.W. Stevens, and J. Montmartin. Bayesian cost–effectiveness analysis from clinical trial data. *Statistics in Medicine*, 20(5):733–753, 2001.

A. O'Hagan, C.E. Buck, A. Daneshkhah, J.R. Eiser, P.H. Garthwaite, D.J. Jenkinson, J.E. Oakley, and T. Rakow. *Uncertain Judgements: Eliciting Experts' Probabilities.* Wiley, New York, 2006.

I. Olkin, A.J. Petkau, and J.V. Zidek. A comparison of n estimators for the binomial distribution. *Journal of the American Statistical Association,* 76(375):637–642, 1981.

S. Palmer, S. Syford, and J. Raftery. Types of economic evaluation. *British Medical Journal,* 318(7194):1349, 1999.

G. Parmigiani. *Modelling in Medical Decision Making: A Bayesian Approach (Statistics in Practice).* Wiley, Chichester, 2002.

D. A. Pettitt, S. Raza, B. Naughton, A. Roscoe, A. Ramakrishnan, A. Ali, B. Davies, S. Dopson, G. Hollander, J. A. Smith, and D. A. Brindley. The limitations of QALY: A literature review. *Journal of Stem Cell Research & Therapy,* 6:334, 2016.

L. Piccinato. A Bayesian property of the likelihood sets. *Statistica,* 44: 197–204, 1984.

J.L Pinto, C. López, X. Badía, A. Corna, and A. Benavides. Cost effectiveness analysis of highly active antiretroviral therapy in HIV asymptomatic patients. *Medicina Clínica,* 114(Suppl 3):62–67, 2000.

J.L Pinto, G. Loomes, and R. Brey. Trying to estimate a monetary value for the QALY. *Journal of Health Economics,* 28(3):553–562, 2009.

J. L. Pinto–Prades, F. I. Sánchez–Martínez, B. Corbacho, and R. Baker. Valuing QALYs at the end of life. *Social Science & Medicine,* 113: 5–14, 2014.

S.J. Pocock, S.E. Assmann, L.E. Enos, and L.E. Kasten. Subgroup analysis, covariate adjustment and baseline comparisons in clinical trial reporting: Current practice and problems. *Statistics in Medicine,* 21(19):2917–2930, 2002.

D. Polsky, H.A. Glick, R. Wilke, and K. Schulman. Confidence intervals for cost–effectiveness ratios: A comparison of four methods. *Health Economics*, 6(3):243–252, 1997.

E. Puigdollers, F. Cots, M.T. Brugal, L. Torralba, and A. Domingo–Salvany. Programas de mantenimiento de metadona con servicios auxiliares: Un estudio de coste–efectividad. *Gaceta Sanitaria*, 17(2): 123–130, 2003.

K. Raikou, A. Gray, A. Briggs, R. Stevens, C. Cull, A. McGuire, P. Fenn, I. Stratton, R. Holman, R. Turner, and UK Prospective Diabetes Study Group. Cost effectiveness analysis of improved blood pressure control in hypertensive patients with type 2 diabetes: UKPCS 40. *British Medical Journal*, 317(7160):720–726, 1998.

G. Richardson and E. Manca. Calculation of quality adjusted life years in the published literature: A review of methodology and transparency. *Health Economics*, 13(12):1203–1210, 2004.

D. Rios and F. Ruggeri, editors. *Robust Bayesian Analysis*, volume 152 of *Lectures Notes in Statistics*. Springer, New York, 2000.

R. Robinson. Cost–benefit analysis. *British Medical Journal*, 307 (6907):793–795, 1993.

V.K. Rohatgi. *An Introduction to Probability Theory and Mathematical Statistics*. Wiley, New York, 1976.

F. Sassi. Calculating QALYs, comparing QALY and DAY calculations. *Health Policy and Planning*, 21(5):402–408, 2006.

L.J. Savage. *The Foundation of Statistics*. Wiley, New York, 1954.

L.J. Savage. Elicitation of personal probabilities and expectations. *Journal of the American Statistical Association*, 66(336):783–801, 1971.

M. Schlander and J. Richardson. The evolving health economics evaluation paradigm and the role of the QALY. *Value in Health*, 12(7): A400, 2009.

J.G. Scott and J.O. Berger. Bayes and empirical–Bayes multiplicity adjustment in the variable–selection problem. *The Annals of Statistics*, 38(5):2587–2619, 2010.

M. Sculpher. Subgroups and heterogeneity in cost–effectiveness analysis. *PharmacoEconomics*, 26(9):799–806, 2008.

M. Sculpher and A. Gafni. Recognizing diversity in public preferences: The use of preference sub–groups in cost–effectiveness analysis. *Health Economics*, 10(4):317–324, 2001.

D.J. Spiegelhalter and L.S. Freedman. A predictive approach to selecting the size of clinical trial based on subjective clinical opinion. *Statistics in Medicine*, 5(1):1–13, 1986.

D.J. Spiegelhalter, L.S. Freedman, and M.K.B. Parmar. Bayesian approaches to randomized trials (with discussion). *Journal of the Royal Statistical Society, Series A (Statistics in Society)*, 157(3):357–416, 1994.

D.J. Spiegelhalter, L.S. Freedman, and M.K.B. Parmar. Bayesian approaches to randomized trials. In D. Berry and D. Stangl, editors, *Bayesian Biostatistics*, pages 67–108. Marcel Dekker, New York, 1995.

D.J. Spiegelhalter, J.P. Myles, D.R. Jones, and K.R. Abrams. Bayesian methods in health technology assessment: A review. *Health Technology Assessment*, 4(38):1–130, 2000.

S. Spiegelhalter, A. Thomas, N. Best, and D. Lunn. OpenBUGS user manual. 2014.

J.W. Stevens and A. O'Hagan. Bayesian methods for cost–effectiveness analysis. *ISPOT News*, 7:303–315, 2001.

A.A. Stinnett and J. Mullahy. Net health benefits: A new framework for the analysis of uncertainty in cost–effectiveness analysis. *Medical Decision Making*, 18(Suppl 2):S68–S80, 1998.

S.D. Sullivan, J. Watkins, B. Sweet, and S.D. Ramsey. Health technology assessment in health–care decisions in the United States. *Value in Health*, 12(Suppl 2):S39–S44, 2009.

A.J. Sutton and J.P. T. Higgins. Recent developments in meta–analysis. *Statistics in Medicine*, 27(5):625–650, 2008.

M. Tambour and N. Zethraeus. Bootstrap confidence intervals for cost–effectiveness ratios: Some simulation results. *Health Economics*, 7(2): 143–147, 1998.

M. Tambour, N. Zethraeus, and M. Johannesson. A note on confidence intervals in cost–effectiveness analysis. *International Journal of Technology Assessment in Health Care*, 14(3):467–471, 1998.

G.W. Torrance. Toward a utility theory foundation for health status index models. *Health Services Research*, 11(4):349–369, 1976.

G.W. Torrance. Measurement of health state utilities for economic appraisal. *Journal of Health Economics*, 5(1):1–30, 1986.

G.W. Torrance. Utility approach to measuring health–related quality of life. *Journal of Chronic Diseases*, 40(6):593–600, 1987.

G.W. Torrance. Designing and conducting cost–utility analyses. In B. Spilker, editor, *Quality of Life and Pharmacoeconomics in Clinical Trials*, pages 1105–1111. Lippincott–Raven, Philadelphia, 1995.

G.W. Torrance and D. Feeny. Utilities and quality–adjusted life years. *International Journal of Technology Assessment in Health Care*, 5 (4):559–575, 1989.

F. Torres–Ruiz, E. Moreno, and F.J. Girón. Intrinsic priors for model comparison in multivariate normal regression. *Revista de la*

Real Academia de Ciencias Exactas, Físicas y Naturales. Serie A. Matemáticas, 105(2):273–289, 2011.

I.S. Udvarhelyi, G.A. Colditz, A. Rai, and A.M. Epstein. Cost–effectiveness and cost–benefit analyses in the medical literature. Are the methods being used correctly? *Annals of Internal Medicine*, 116 (3):238–244, 1992.

B.A. Van Hout, M.J. Al, G.S. Gordon, and F.F. Rutten. Costs, effects and c/e–ratios alongside a clinical trial. *Health Economics*, 3(5): 309–319, 1994.

D.J. Vanness and W.R. Kim. Bayesian estimation, simulation and uncertainty analysis: The cost effectiveness of Ganciclovir Prophylaxis in liver transplantation. *Health Economics*, 11(6):551–566, 2002.

F.J. Vázquez–Polo and M.A. Negrín. Incorporating patients' characteristics in cost–effectiveness studies with clinical trial data: A flexible Bayesian approach. *Statistics and Operation Research Transaction*, 28(1):87–108, 2004.

F.J. Vázquez–Polo, M.A. Negrín, X. Badía, and M. Roset. Bayesian regression models for cost–effectiveness analysis. *The European Journal of Health Economics*, 6(1):45–52, 2004.

F.J. Vázquez–Polo, M.A. Negrín, J. Cabasés, E. Sánchez, J.M. Haro, and L. Salvador–Carulla. An analysis of the costs of treating schizophrenia in Spain: A hierarchical Bayesian approach. *The Journal of Mental Health Policy and Economics*, 8(3):153–165, 2005a.

F.J. Vázquez–Polo, M.A. Negrín, and B. González. Using covariates to reduce uncertainty in the economic evaluation of clinical trial data. *Health Economics*, 14(6):545–557, 2005b.

J. von Neumann and O. Morgenstern. *Theory of Games and Economic Behavior*. Princeton University Press, Princeton, 1944.

P. Wakker and M.P. Klaasen. Confidence intervals for cost–effectiveness ratios. *Health Economics*, 4(5):373–381, 1995.

A. Wald. *Statistical Decision Functions*. Chelsea Publishing Company, New York, 2nd edition, 1971.

L.A. Wasserman. A robust Bayesian interpretation of likelihood regions. *The Annals of Statistics*, 17(3):1387–1393, 1989.

M.C. Weinstein. Principles of cost–effective resource allocation in health care organizations. *International Journal of Technology Assessment in Health Care*, 6(1):93–103, 1990.

M.C. Weinstein and W.B. Stason. Foundations of cost–effectiveness analysis for health and medical practices. *New England Journal of Medicine*, 296(13):716–721, 1977.

A.R. Willan and A.H. Briggs. *Statistical Analysis of Cost–Effectiveness Data*. Wiley, Chichester, UK, 2006.

A.R. Willan and M.E. Kowgier. Cost–effectiveness analysis of a multinational RCT with a binary measure of effectiveness and an interacting covariate. *Health Economics*, 17(7):777–791, 2008.

A.R. Willan and B. O'Brien. Cost prediction models for the comparison of two groups. *Health Economics*, 10(4):363–366, 2001.

A.R. Willan and B.J. O'Brien. Confidence intervals for cost–effectiveness ratios: An application of Fieller's theorem. *Health Economics*, 5(4):297–305, 1996.

A.R. Willan, A.H. Briggs, and J.S. Hoch. Regression methods for covariate adjustment and subgroup analysis for non–censored cost–effectiveness data. *Health Economics*, 13(5):461–475, 2004.

R. Willke, H.A. Glick, D. Polsky, and K. Schulman. Estimating country–specific cost–effectiveness from multinational clinical trials. *Health Economics*, 7(6):481–493, 1998.

R.L. Wolpert. Eliciting and combining subjective judgements about uncertainty. *International Journal of Technology Assessment in Health Care*, 5(4):537–557, 1989.

B.T. Yates. Cost–effectiveness analysis of a residential treatment center for disturbed preadolescents. In G. Landsberg, W.D. Neigher, R.J. Hammer, C. Windle, and R. Woy, editors, *Evaluation in Practice: A Sourcebook of Program Evaluation Studies from Mental Health Care Systems in the United States*. National Institute of Mental Health, Washington, D.C., 1979.

B.T. Yates, D.L. Lockwood, K. Saylor, M. Andrews, S. Shiraki, P. Delany, K. Bestamen, J. Heath, J.A. Inciardi, J. Merrill, J.K. Myers, and G. Zarkin. *Measuring and Improving Cost, Cost–Effectiveness and Cost–Benefit for Substance Abuse Treatment Programs*. National Institutes of Health, U.S., 1999.

S. Zacks. *Parametric Statistical Inference: Basic Theory and Modern Approaches*. Pergamon Press, 1981.

P.W. Zehna. Invariance of the maximum likelihood estimators. *The Annals of Mathematical Statistics*, 37(3):744, 1966.

A. Zellner. On assessing prior distributions and Bayesian regression analysis with g–prior distributions. In P. Goel and A. Zellner, editors, *Bayesian Inference and Decision Techniques: Essays in Honor of Bruno de Finetti*, pages 233–243. Elsevier Science Publishers, Inc, New York, 1986.

A. Zellner and A. Siow. Posterior odds ratios for selected regression hypotheses. *Trabajos de Estadística y de Investigación Operativa*, 31 (1):585–603, 1980.

Index

Printed and bound by CPI Group (UK) Ltd, Croydon, CR0 4YY
24/10/2024
01778281-0005